Williar

ON THE

POWER WISDOM AND GOODNESS OF GOD

AS MANIFESTED

OF ANIMALS AND

AND INSTINCTS

Volume

Elibron Classics
www.elibron.com

Elibron Classics series.

© 2005 Adamant Media Corporation.

ISBN 1-4021-7155-2 (paperback)
ISBN 1-4212-8398-0 (hardcover)

This Elibron Classics Replica Edition is an unabridged facsimile
of the edition published in 1835 by William Pickering, London.

Elibron and Elibron Classics are trademarks of
Adamant Media Corporation. All rights reserved.

This book is an accurate reproduction of the original. Any marks, names, colophons, imprints, logos or other symbols or identifiers that appear on or in this book, except for those of Adamant Media Corporation and BookSurge, LLC, are used only for historical reference and accuracy and are not meant to designate origin or imply any sponsorship by or license from any third party.

# THE BRIDGEWATER TREATISES

## ON THE POWER WISDOM AND GOODNESS OF GOD

## AS MANIFESTED IN THE CREATION

———

### TREATISE VII

ON THE HISTORY HABITS AND INSTINCTS OF ANIMALS

BY THE REV. WILLIAM KIRBY, M.A.

IN TWO VOLUMES

VOL II

*" C'EST, LA BIBLE A LA MAIN, QUE NOUS DEVONS ENTRER DANS LE TEMPLE AUGUSTE DE LA NATURE, POUR BIEN COMPRENDRE LA VOIX DU CRÉATEUR."*

GAEDE.

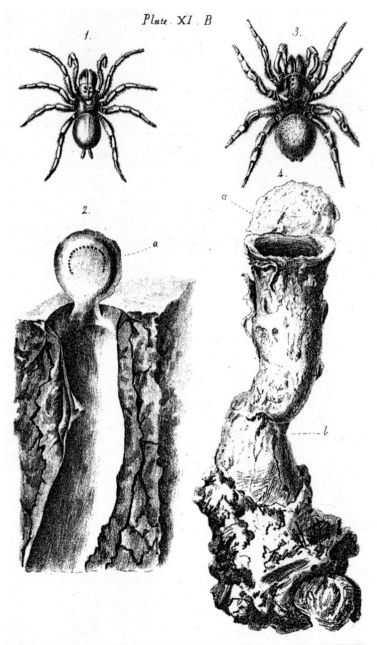

Plate. XI. B

C.M.Curtis del.          Meijer lith.

ON THE

# POWER WISDOM AND GOODNESS OF GOD

## AS MANIFESTED IN THE CREATION
## OF ANIMALS AND IN THEIR HISTORY HABITS
## AND INSTINCTS

BY THE

### REV. WILLIAM KIRBY, M.A. F.R.S. ETC.

RECTOR OF BARHAM.

### VOL II

ALDI

DISCIP.

ANGLVS

## LONDON

### WILLIAM PICKERING

1835

C. WHITTINGHAM, TOOKS COURT, CHANCERY LANE.

# CONTENTS

## OF THE SECOND VOLUME.

# EXPLANATION OF THE PLATES.

## VOLUME II.

# HISTORY, HABITS, AND INSTINCTS

# OF ANIMALS.

---

## CHAPTER XIII.

*Functions and Instincts. Cirripedes and Crinoïdeans.*

### CIRRIPEDES.

THERE is a class of animals defended by multi-valve shells, separated from the Molluscans not only by the more complex structure of their shells, but also by very material differences in the organization of the creatures that inhabit them. These Linné considered as forming a single genus, which he named *Lepas*, a word derived from the Greek lexicographers, and explained by Hesychius as meaning a kind of shell-fish that adheres to the rocks. In this country these animals are known by the general name of *Barnacles*. Lamarck I believe, was the first who regarded them as entitled to the rank of a class, which he denominated *Cirr-hipeda*, not conscious, that by the insertion of the aspirate, he made his term, like *Monoculus*, half Greek and half Latin : later writers who have adopted the class, to avoid this barbarism, have changed the term to *Cirrhopoda*, but as this gives a different meaning to the word, changing *fringed* or *tendril-legs*,[1] very happily expressing the most striking character of the animals intended, into *yellow-legs*,[2] which does not indicate any prominent feature, I shall, after Dr. Leach and Mr. W. S. Mac Leay, omitting the aspirate, call them *Cirripeda*, or Cirripedes.

These animals have a soft body, protected by a multivalve shell. They are without eyes, or any distinct head ; have no powers of locomotion, but are fixed to various substances. Their body, which has no articulations, is enveloped in a kind of mantle, and has numerous tentacular arms, consisting of many joints, fringed on each side, and issuing by pairs from jointed pedicles : their mouth is armed with transverse toothed jaws in pairs, which, like the mandibles of the Crustaceans, are furnished with a feeler ; they have a knotty longitudinal spinal chord ; gills for respiration ; and for circulation, a heart and vascular system.

---

[1] Lat. *Cirri.*          [2] Gr. κιρρος.

This class is divided into two Orders.

1. The *first* consists of the *Lepadites,* or Goose-barnacles,[1] the species of which are distinguished by a tendinous, contractile, and often long tube, fixed by its base to some solid marine substance, supporting a compressed shell, consisting of valves united to each other by membrane, and by having six pairs of tentacular arms. They are usually found in places exposed to the fluctuations of the waves. One genus[2] appears to perforate rocks to form a habitation. These animals roll up and unroll their arms with great velocity, thus creating a little whirlpool, that brings to their mouth an abundant supply of animalcules, an action which Poli compares to fishermen casting a net. Some species, instead of shell, are covered by a membranous sac, having occasionally very minute shelly valves.[3]

2. The *second* Order of Cirripedes consists of the *Balanites,* or Acorn-barnacles, which are distinguished from the Lepadites by a shelly, instead of a tendinous tube, the mouth of which is closed by an operculum, usually consisting of four valves. The animals of this Order are commonly regarded as sessile; but, if Lamarck is right in considering the valves of the shell of the Lepadites as analogous to the operculum of

---

[1] *Anatifa. Pentelasmis,* &c.   [2] *Lithotrya.*
[3] *Anatifa coriacea et leporina.*

the Balanites, as it seems to be, and their tendinous tube as really a part of the body of the animal—as its being organized, living, and muscular, seems to prove—then it must be analogous to the shelly tube of the latter, and both must be considered as elevated by a footstalk. This tube, in the Balanites, consists usually of six pieces, soldered, as it were, together; and in several species, as in the common sea-acorn,[1] of a triangular shape, and having their acute angle alternately at the base and at the mouth of the tube. The base of the tube generally takes the form of the bodies upon which it is fixed, and is sometimes composed of shell, sometimes of membrane, and sometimes it is incomplete. The animal, in this Order, has twenty-four tentacular arms, shorter than those of the Lepadites, consisting of two sorts, namely, six pairs of large similar ones, but unequal in size, placed above; and as many smaller pairs, dissimilar and unequal, and placed below. One pair of these is much larger than the others. In the water they keep these tentacles[2] in perpetual motion, and thus arrest, or, by producing a current to their mouth, absorb the animalcules, which constitute their food. They not only fix

---

[1] *Balanus Tintinnabulum.*

[2] These organs, though called tentacles, from their use, seem rather analogous to the *antennæ* and other *jointed* organs of *Condylopes.*

themselves upon inanimate substances, such as rocks, stones, the hulls of ships, &c. but also upon various marine animals and plants. Thus some are found on Zoophytes, as sponges and madrepores; others attached closely to each other on shell-fish, especially bivalves, so closely that the point of a pin cannot be thrust between them. One species takes its station on the shell of the turtle;[1] others plant themselves in the flesh of the seal; and others bury their tube in the unctuous blubber of the whale.

If we compare the animals of the above Orders with each other, we shall find that they are fitted by their Creator to collect their food in different ways. The *Lepadites*, by means of their long contractile flexible tube, can rise or sink, and bend themselves in different directions, so as, in some sort, to pursue their prey; their tentacles, also, from their greater length, seem to further this end: these, according to Poli's metaphor above alluded to, they can throw out and draw in laden with fry, as a fisherman does his net. When their prey is in their mouth, it is subjected to the action of their toothed jaws, which seem more numerous and powerful than those of the Balanites; and as the valves forming the shell are more numerous and connected by membrane, and the whole shell more compressed than the operculum of the last named animals, we may

[1] *Coronula testudinaria.*

suppose that they are capable of a more varied action, and one that may perhaps add to the momentum of the masticating organs. Hence we may conjecture that the animals destined to form their nutriment, may be larger, so as to require more exertion and force, both to take and to masticate.

In the other Order, the structure of the *Balanites* seems to indicate merely the protrusion and employment of their tentacles; and being usually attached to floating bodies, such as the hulls of ships, or parasitic upon locomotive animals, riding as they do upon the back of the turtle, the dolphin, and the whale, they may visit various seas in security, and feast all the while, with little trouble and exertion, upon animalcules of every description, the produce of arctic, temperate, and tropical seas.

With respect to their place in nature, it seems not quite clear whether they should be regarded as leading from the Molluscans, with which Cuvier arranges them, towards the Crustaceans, and they certainly seem to have organs borrowed from both; their shells and mantle in some degree from one, and their palpigerous mandibles and jointed organs, proceeding in pairs from a common footstalk—like the interior antennæ of the lobster—and knotty spinal chord from the other: but with respect to their jointed organs, I must observe that they still more closely resemble those of some of the Encrinites,[1] like them being fringed on each side, though not with organs of that description. A learned naturalist, Mr. W. S. Mac Leay, is of opinion that the Echinidans, or sea urchins, exhibit some approximation to the Balanites.[2]  If, indeed, we compare the genus *Coronula* with an *Echinus*, we shall discover several points in which their structure agrees.  We learn from Lamarck, that the pieces of the so called operculum, which close the mouth of the former shell, are affixed rather to the animal than to the shell.  Thus the operculum, in some sort, represents the jaws of an Echinus, though consisting of fewer pieces, and the tube appears divided into alleys, like the crust of that animal. These circumstances seem to prove some affinity between the Cirripedes and Radiaries ; they appear also to have some points in common with Savigny's Nereïdeans, especially *Amphitrite*.[3] Weighing all these circumstances, I have thought it best to place the Cirripedes immediately before the Entomostracan Crustaceans.

But what if these Cirripedes should at last prove to be, not the guides to the great Crustacean host, but its legitimate progeny ?  This has been asserted, at least partially, by a modern zoologist, who has assigned his reasons for this

[1] See PLATE III. B. FIG. 1.
[2] *Hor. Ent.* i. 312.          [3] *Ibid.*

singular and startling opinion. I will not say
the thing is impossible—for with God all things
are possible—but it certainly appears in the
highest degree improbable. That a *Zoea* should
become a crab is sufficiently extraordinary, and
an opinion, as Latreille remarks, which, if it be
not erroneous, has great need of support from
experiment :[1] but that a locomotive animal,
gifted with eyes and legs, should, by an extraordinary metamorphosis, in its perfect state, become a barnacle, without head, eyes, or locomotive organs, can never be admitted till confirmed
by repeated experiments of the most able and
practised zoologists, so as to place the matter
beyond dispute. I by no means, however, mean
to assert that Mr. Thompson did not think he
saw what he has stated, in both cases, to take
place, but he was probably deceived by appear-
ances in some such way as he states Slabber to
have been.[2]

A single fact, observed by Poli, is sufficient to
overturn this whole hypothesis. This illustrious
conchologist relates that he had an opportunity
of examining the immense fecundity of the
sessile barnacles. " In the beginning of June
he found innumerable aggregations of them,
covering certain boats that had been long stationary, which, when closely examined, were so

[1]   *Cours D'Entomologie*, i. 385.
[2]   *Zool. Research.* No. i. 7.

minute, that single shells were not bigger than the point of a needle; and that from that time they grew very rapidly, and arrived at their full size in October." These very minute ones must have been hatched from the egg, and not produced from larves.

With regard to the functions and instincts of these Cirripedes, very little has been observed. We see from the above account of them, that, like many other animals amongst the lowest grades of the animal kingdom, they are furnished with particular organs adapted to the capture of animalcules and other minor inhabitants of the deep, which they help to keep within due limits. Probably they act upon the substances to which they attach themselves, and promote the decomposition of shells, and other exuviæ of defunct animals, and also of the rocks and ligneous substances on which they take their station. Of this we are sure, that they work His work who gave them being, and assigned them their several stations in the world of waters.

### CRINOÏDEANS.

In the deepest abysses of the ocean, it is probable, lurks a tribe of plant-like animals, to judge from its numerous fossil remains, abounding in genera and species that are very rarely

seen in a recent state, and which, from a supposed resemblance between the prehensory
organs or arms, surrounding the head or mouth of
several species belonging to the tribe, when their
extremities converge, to the blossom of a liliaceous
plant, have been denominated *Encrinites* and
*Crinoïdeans*.[1]   It was not my original intention,
as little or nothing was known with respect to
the habits and station of the few recent ones that
have been met with—except that one has been
taken in the seas of Europe, and three in the
West Indies, namely, near Martinique, Bar-
bados, and Nevis—to have introduced them into
the present work, but having subsequently seen
fragments of a specimen, taken either in the
Atlantic or Pacific, I am not certain which, and
upon examining it under the microscope, finding
evident traces of suckers on the underside of
its fingers, and of the tentacles that form its
fringes,[2] a circumstance I found afterwards mentioned by Ellis, and which throws some light
upon their economy, I felt that I ought not to
pass them wholly without notice, and finding in
the Hunterian Museum a very fine specimen
which does not appear to have been figured, for
the figure given by Ellis seems to have been
taken from Dr. Hunter's specimen, now at Glas-
gow, and Mr. Miller's from a specimen of Mr.
Tobin's, now in the British Museum, by the kind

---

[1] From κρινον, a lily.        [2] PLATE III. B. FIG. 2.

permission of the Curators of the Museum in Lincoln's Inn Fields, I was allowed to have a figure of it taken by my artist, Mr. C. M. Curtis.[1]

Lamarck has placed the Crinoïdeans, led probably by their plant-like aspect, in the same Order with his *Floating Polypes*,[2] not aware that the majority are evidently *fixed*, but Cuvier and most modern zoologists consider them, with more reason, as forming a family of the *Stelleridans*, from which the way to them is by the genus *Comatula*, remarkable for its jointed rays fringed on each side. The *Marsupites*, as Mr. Mantell, after Mr. Miller, has observed, form the link which connects the proper or pedunculated Crinoïdeans with the Stelleridans. If we compare them again with the class last described, the *Cirripedes*, especially the *Lepadites*, we shall find several points which they possess in common. In the first place both sit upon a footstalk, though of a different structure and substance; the animal in both, in its principal seat, is protected by shelly pieces or valves; the head or mouth in both, is surrounded by dichotomizing articulated organs, involuted, and often converging at the summit, and fringed on each side, in the Crinoïdeans, with a series of lesser digitations, and in the Cirripedes with a dense fringe of hairs. If the opinion of Mr. W. S. Mac Leay, stated above,

---

[1] PLATE III. B. FIG. 1.    [2] *Polypi natantes.*

that some of the *Echinoderms* exhibit an approximation to some of the *Cirripedes*, is correct, as it seems to be, the *Crinoïdeans*, though still far removed, would form one of the links that concatenate them ; or if their connection is thought merely analogical, the *Balanites* would be the analogues of the *Echinidans* and of the sessile Crinoïdeans, and the Lepadites of the pedunculated ones.

The following characters distinguish the *Pentacrinites*, to which Tribe all the known recent species belong.

*Animal,* consisting of an angular flexible column, composed of numerous joints, articulating by means of cartilage, and perforated for the transmission of a siphon or intestinal canal, and sending forth at intervals, in whorls, several articulated cylindrical branches, curving into a hook at their summit ; fixed at its base, and supporting at its free extremity a cup-like body, containing the mouth and larger viscera, consisting of several pieces, terminating above in five (or six) dichotomizing, articulated, semi-cylindrical arms, fringed with a double series of tentacular jointed digitations, furnished below on each side with a series of minute suckers : these arms, when expanded, resemble a star of five (or six) rays, and when they converge, a pentapetalous or hexapetalous liliaceous flower. The whole animal, when alive, is supposed to be invested with a gelatinous muscular integument.

In the specimen figured by Mr Ellis, and that in the Hunterian Museum, there appear to be *six* arms springing from the so-called pelvis, but the natural number appears to be *five*, corresponding with the pentagonal column. Mr. Miller seems to be of opinion that the species described by M. Guettard, and that which he has himself figured, are the same species, and synonymous with the *Isis Asteria* of Linné and the *Encrinus Caput Medusæ* of Lamarck, but to judge from the figures of the first in Parkinson,[1] and of the other in Miller,[2] compared with that which is given in this work,[3] the last seems to differ from both, as well in the pelvis, as in the dichotomies, and length of the arms; its suckers likewise appear to be circular,[4] and not angular as they are described by Mr. Miller under the name of plates.[5] If this observation turns out correct, I would distinguish the last species by the name of *Pentacrinus Asteria*.

The stem of the Crinoïdeans consists of numerous joints, united by cartilages, which exhibit several peculiarities; in the first place the upper and under side is beautifully sculptured, so as to represent a star of five rays, or a pentapetalous

---

[1] *Organic Remains*, ii. *t.* xix. *f.* 1.   [2] *Crinoïdea*, 48. *t.* 1.
[3] PLATE III. B. FIG. 1.   [4] *Ibid.* FIG. 2.
[5] *Ubi supr.* 54. *t.* ii. *f.* 6.

flower; the Creator's object in this structure appears to be the attachment of the cartilage that connects them, and, perhaps, to afford means for a degree of rotatory motion, as well as to prevent dislocations, and also to increase the flexure of the stem according to circumstances, and the will of the animal. For the transmission of the siphon, whether a spinal chord, or intestinal canal, or both, each joint of the column is perforated, the aperture being round in some, and floriform in others. The whole stem, with its whorls of branches, exhibits a striking resemblance to the branch of the common horse-tail.[1] The entire structure seems calculated to enable the animal to bend its stem, which appears very long, in any direction, like the Lepadites, and thus as it were to pursue its prey; we may suppose that the branching arms, fingers, and their lateral organs, when they are extended horizontally and all expanded, must form an ample net, far exceeding that of the Cirripedes, which, when they have their prey within its circumference, by converging their arms, and closing all their digitations, and employing their suckers, they can easily so manage as to prevent the escape of any animal included within the meshes of their net.

With regard to their functions, and what ani-

---

[1] *Equisetum arvense.*

mals their Creator has given a charge to them
to keep within due limits, little can be known by
observation; as nothing like jaws has been discovered in them, in which they differ from the
Cirripedes, it should seem that either their food
must consist of animalcules that require no mastication, or, if they entrap larger animals, that
they must suck their juices, which seems to be
Mr. Miller's opinion.[1]   This idea is rendered not
improbable by the vast number of suckers by
which their fingers, and their lateral branches
or tentacles as they are called, are furnished; by
these they can lay fast hold of any animal too
powerful to be detained in their net by any
other means, and subject it to the action of their
proboscis.

From the great rarity of recent species of
these animals, it should seem that the metropolis
of their race is in the deepest abysses of the
world of waters.   "It appears," says Bosc,[2]
"that the species were extremely numerous in
the ancient world, perhaps, those actually in
existence are equally so, for I suspect that all
inhabit the depths of the ocean, a place in which
they may remain to eternity without being
known to man.".

Naturalists very often, too hastily, regard
species as extinct, that are now found only in

---

[1] *Crinoïdea*, 54.      [2] *N. D. D'Hist. Nat.* x. 224.

a fossil state, forgetting that there may be many stations fitted for animal or vegetable life, that are still, and, perhaps, always will be inaccessible to the investigator of the works of the Creator, where those mourned over as for ever lost, may be flourishing in health and vigour.

## CHAPTER XIV.

### *Entomostracan Condylopes.*

WE are now arrived at a great branch of the animal kingdom, which, in its higher tribes, exhibits Divine Wisdom, acting, in and by the instincts of creatures, small indeed in bulk, but mighty in operation, in a way truly admirable, indicating, in a most striking manner, the source from which it proceeds.

Some modern zoologists do not regard this vast and interesting branch as forming a group by itself, but have associated with it, under a common name, several of the preceding classes. Carus, in his *Class of Articulated Animals*,[1] includes Lamarck's *Worms* and *Annelidans;* and Dr. Grant, in his *Sub-kingdom*, bearing the same appellation, adds to these the *Wheel-animalcules*,[2] and *Cirripedes*.[3]

[1] *Articulata.*      [2] *Rotifera.*      [3] *Cirrhopoda.*

I cannot help thinking, however—taking the whole of their organization and structure into consideration, particularly their powers and means of locomotion and prehension—that it is best to regard those animals having *jointed legs*, and, mostly, *a body formed of two or more segments*, as constituting a separate Sub-kingdom. This is the view that my late illustrious and lamented friend, Latreille, has taken of this great group, named by him, from the above circumstance, *Condylopes*,[1] which term, since that of *Annulose animals*,[2] sometimes used, is synonymous with *Annelidans*, I shall adopt in the present work.

The distinctive characters of this great group, or Sub-kingdom, may be given in few words:

ANIMAL, not fixed by its base, but locomotive.

*Body*, in the great majority, consisting of two or more segments.

*Legs*, jointed.

The first of these characters distinguishes the *Condylopes* from the last class, the *Cirripedes*, which are fixed by their base, whereas the present tribe are more free in their motions than most of the animals of the preceding groups; and the two last from the *Annelidans*, which, though annulated, are not insected, and have no jointed legs.

---

[1] *Condylopa*, from κονδυλοι, joints, and πους, a foot.
[2] *Annulosa*.

Cuvier, Latreille, and most other zoologists, consider this section of the animal kingdom as subdivided into *three* great Classes—*Crustaceans, Arachnidans,* and *Insects*: Dr. Leach, taking the respiratory organs for his guide, also begins with three *primary* Sections, those, namely, which have *gills*, those which have *sacs*, and those which have *tracheæ*, for respiration ; and out of these he forms *five* Classes, viz. *Crustaceans, Arachnoidans, Acarines, Myriapods,* and *Insects.* The first and last of these Classes he further subdivides, each into two Sub-classes : the Crustaceans into *Entomostracans* and *Malacostracans ;* and Insects into *Ametabolians* and *Metabolians,* or those that do not undergo a metamorphosis, and those that do. So that according to his *primary* Section his system is *ternary ;* according to his *secondary* it is *quinary ;* and according to his *tertiary* it is *septenary.* I shall mostly follow him in each of these last subdivisions.

Having made these remarks upon the Condylopes in general, I must now proceed to one of the Classes above enumerated : but here, at first, it seems difficult to ascertain which ought to be regarded as forming the first step in an ascending series,—a difficulty, indeed, which often arrests the course of the student of the works of his Creator, for, when any one, in a philosophic spirit, after a careful survey, sits down to trace the paths by which Divine Wisdom seems to have passed in the creation, and the arrangement and connection of the various groups of organized beings, he is lost and bewildered in a most intricate and mazy labyrinth, in which paths intersect each other at every angle, and when he thinks he is travelling in a straight road he often comes to branches leading off from it, which render it uncertain in which direction he ought to proceed, in order best to attain the object he is pursuing.

Such indeed is the perplexity of animated nature, that it is impossible to see clearly the arrangement of the objects that constitute either the vegetable or the animal kingdom; and in order to get any tolerable notion of them, as God has placed them, when we have reached a certain station we are often obliged to retrograde, and begin a branch, from the point of its divergement, far removed from that to which we have arrived.

Latreille, in the last edition of the *Règne Animal*, divides his *Crustaceans* into *two* Sub-classes, the first of which, after Aristotle, he denominates *Malacostracans;*[1] and the second, after Müller, *Entomostracans:*[2] these, on account of a connection which seems to exist between them and the *King-crab,*[3] he places immediately before the *Arachnidans.* I agree with this learned entomo-

---

[1] *Malacostraca.*　　　　[2] *Entomostraca.*
[3] *Limulus Polyphemus.*

logist, in considering them as inferior to the
proper Crustaceans, and shall therefore begin
the Condylope group with some account of
them. Like the infusory animalcules, they
form a kind of centre, sending forth rays to different points, some inclosed in a bivalve shell,
seeming to tend towards the *Molluscans*;[1] others
assuming more of the *Crustacean* form;[2] a third
looking to the *Arachnidans*;[3] and a fourth to the
*Thysanuran*, or Sugar-louse tribe;[4] with other
forms that might be enumerated, some of which
are perfectly anomalous, so that it appears almost
indifferent where they are placed. As there is,
however, evidently some affinity between the Entomostracans and the Cirripedes, not only in both
being furnished with jointed organs for their
motions, but also in some of the former being
inclosed in shells, and in others by the brisk
agitation of their legs, producing a current in
the water to their mouths, as De Geer states of
the Water-flea:[5] this furnishes a further argument for placing them next to the latter tribe.

It is difficult, and next to impossible, to fix
upon any characters that are common to the
whole of this remarkable Class. Generally
speaking, but not invariably, they are covered,
not by a calcareous and solid, but by a horny
and thin integument. They vary considerably

---

[1] *Cypris*, &c.    [2] *Branchipus*.    [3] *Limulus*.
[4] *Cyclops*.    [5] *Daphnia Pulex*. *De Geer*, vii. 453.

in the number of their antennæ and legs, the former often branching, and used as oars, and the latter usually being connected with their respiration, evincing the analogy between these legs and the ciliæ of the Rotatories, and tentacles of the Polypes ;[1] in the majority these organs are not calculated for prehension. One group of them lives by suction, and is parasitic upon other aquatic animals: the great body, however, masticate their food, but without the aid of maxillary legs. Their eyes are generally sessile, and a considerable number of them have only one, or rather two eyes enveloped by a common cornea.[2]

Latreille, in his *Cours D'Entomologie*, divides this Class—regarded by Linné as forming one genus, which he named *Monoculus*—into *six* Orders; but it will be sufficient here to adopt his division of them in the *Règne Animal*, into *two*, which, as separating the fresh-water from the marine genera, is more simple, and better suited to my purpose. These Orders he names Branchiopods and Pœcilopods.

1. The *Branchiopods* are all very minute, and several of them microscopic animals. Their mouth consists of an upper lip, two mandibles, a tongue, and one or two pairs of maxillæ. Their legs are natatory, connected with their respira-

---

[1] See above, VOL. I. p. 154, 164.  [2] Roget, *B. T.* ii. 493.

tion—whence their name of *Branchiopods*, or gill-bearing legs—often branching, varying in number from six to more than a hundred.

2. The *Pœcilopods* differ from the preceding Order by the different structure and uses of their legs, which are not branching, and all of them in some, and part of them in others, are prehensory and ambulatory, in some part are also branchial and natatory. They differ likewise by not having the ordinary mandibles and maxillæ, which are sometimes replaced by the spiny hips of the six first pairs of legs, and, in one tribe, by a mouth and oral organs proper for suction.

There is a tribe of parasitic animals, which neither Cuvier nor Latreille have included amongst the Entomostracans, but which Audoin and Milne Edwards conjecture are of a *Crustacean* type. I am speaking of the *Lerneans* of the author first mentioned, which he has placed, but not without hesitation, in his first order,[1] of Intestinal Worms.[2] Dr. Nordmann, however, has made it evident that they undergo a metamorphosis little differing from that of the first Order of the Entomostracans, the Branchiopods, especially *Cyclops;* and he is of opinion, that, in a system, they would follow that genus. Their resemblance is indeed striking in their prepa-

---

[1] *Intestinaux cavitaires.*      [2] *Entozoa,* Rud.

ratory states, but in their last or perfect state, they differ, and like the Pœcilopods, are parasitic; many of them are furnished with a very conspicuous organ, which I shall afterwards describe, for fixing themselves; and their form is very different, their body consisting of two segments, like that of the Arachnidans,[1] though attached to their abdomen, like many of the Branchiopods, they have two egg-pouches.[2] In fact the Lerneans seem scarcely more anomalous amongst the Entomostracans, than the King-crab, and other Pœcilopods. All things considered, perhaps, they may be regarded as forming an osculant group between the two Orders.

The animals of the first Order mostly frequent stagnant waters, moving about with great rapidity. They are generally regarded as predaceous, and are stated to make the infusory animalcules their prey, but some are supposed to be herbivorous, and they abound particularly in waters in which plants are vegetating. As the places that they frequent are very subject to be dried up in the summer-time, it seems probable that a kind Providence has fitted them for this event, by giving them, as well as the Infusories, powers of reviviscence. Latreille thinks that those of them, which, for the protection of

---

[1] PLATE IX. FIG. 5.        [2] *Ibid. f. f.*

their slender and frail branching antennæ and legs, are enclosed in shells, have the power, after drawing in all their organs, of hermetically sealing their shells till the return of moisture.

These little animals differ from the Molluscans, and the other preceding Classes, by the changes of their integument; they do not, like them, when their advance in growth requires it, add to their shells; but, fixing themselves to some substance at hand, they move their limbs, and the valves of their old shells, new ones being already formed underneath, and thus loosening their exuviæ, in a short time they cast those of the whole body; of all their limbs, hairs, plumes, even those that are invisible to the naked eye. Amongst these exuviæ may be detected, not merely the cast skin of the external parts, but that of the internal also. These moults follow each other at an interval of five or six days, and it is not till after the third that the animal has acquired the reproductive faculty.

In the antecedent classes of the animal kingdom, which were almost all inhabitants of the water, we have seen no instances of animals casting their skins, or undergoing any metamorphosis—either in the number or form of their parts—in their progress to their adult state. Some few shell-fish, indeed, are stated to cast their shells, and form others,[1] but a degree of

[1] See above, Vol. I. p. 300.

doubt rests upon the fact. In the Branchiopods, however, a kind of metamorphosis, as well as the moult just described, has long been noticed and recorded.

The young ones of the *Cyclops*, the animal before mentioned as an analogue of the sugar-louse, when first hatched have only *four* legs, their body is nearly round, and has no tail, which led Müller to mistake them for species of a different genus;[1] soon afterwards another pair is acquired, which the same author regarded as a second genus,[2] and so it proceeds till it assumes the perfect form of its kind. Nordmann has given figures of a very remarkable Lernean parasite,[3] which infests the perch, representing its whole progress, from the egg to the perfect insect,[4] which, like the Cyclops, does not acquire all its organs, except at its last metamorphosis.

Our progress upwards, as far as we have at present proceeded, has been a gradual advance, form after form appearing upon the stage of animal existence, each distinguished by characters indicating an elevation as to rank and station. But in the animals amongst which the law in question obtains, we see the same individual, at different periods of its existence, assuming a higher tone of character, and often endued

---

[1] *Amymone.*    [2] *Nauplius.*    [3] *Actheres Percarum.*
[4] PLATE IX. *Egg*, FIG. 1, 2. *Larva*, FIG. 3. *Pupa*, FIG. 4. *Imago*, FIG. 5.

with organs that fit it for a more extended range. Sometimes from being purely aquatic, it becomes a denizen of the earth and the air—or of earth, air, and water at once—and, with this change of character and organs, its Creator wills it to undertake a new charge in the general arrangement of functions and duties.

It will be recollected that a very considerable portion of the food of the higher creatures, especially the birds, is derived from animals that undergo a metamorphosis ; and, that the majority of these in their first state, are more bulky, and contain more nutritive substance than they do when arrived at their last, and, therefore, even in this view, circumstances important to the general welfare may arise from this disposition, and variety of food may also be produced, and more enjoyment to the various animals who are destined to live by the myriad forms of the insect world.

Whether the higher Orders of Crustaceans undergo a real metamorphosis has not been satisfactorily proved.   They are known to change their shells annually, but it has not been observed that this moult is attended by any change of form, or by the acquisition of new locomotive or other organs.   Insects, we know, after their last change do not increase in size ; the Crustaceans are found, however, to vary very much in this respect.   Whether a different law obtains amongst them, from what takes place in insects, and they follow the Batrachian reptiles, which, after they have exchanged the tadpole for the frog, grow till they have arrived at the standard of their respective species, I cannot certainly affirm ; but reasoning from analogy, it seems more probable that the crustaceans should follow the law of animals most nearly related to them, and belonging to the same primary group, than that they should copy the reptiles, animals far removed from them, and of a completely different organization.

There is another point in which this subject of animal metamorphoses may be viewed. Do not these successive changes in the outward form, functions, and locomotions of so many animals, preach a doctrine to the attentive and duly impressed student of animal forms, and their history—do they not symbolically declare to him, that the same individual may be clothed with different forms, in different states of existence, that he may be advanced, after certain preparatory changes, and an intermediate interval of rest and repose, to a much more exalted rank ; with organs, whether sensiferous or locomotive, of a much wider range; with tastes more refined ; with an intellect more developed, and employed upon higher objects; with affections more spiritualized, and further removed from gross matter ?

The multiplication of these creatures, which, like the *Aphides,* are oviparous at one time, and viviparous at another, is sometimes prodigious, and only exceeded by that of the Infusories. A female *Cyclops,* the animal before alluded to, in the space of three months, after one fecundation which serves for several successive generations, lays her eggs ten times, and it has been calculated that from only eight of these ovipositions, allowing forty for each, she might be the progenitrix, incredible as it may seem, of four milliards and a half, or four thousand five hundred millions!!![1] Another animal belonging to a genus of the present order,[2] was observed by Captain Kotzebue in such myriads that the sea exhibited a red stripe, a mile long, and a fathom broad, produced by a species, individually viewed, scarcely visible to the naked eye. How astonishing is the reflection, that in so short a space, in the case of the Cyclops, a single individual should be gifted by its Creator to fill the waters with myriads of animated beings, supposing a single impregnated female at first to have been the surviving inhabitant of any given pool or ditch. Conjecture is lost when we meditate upon the mysterious subject. How can life, as originally imparted, at the interval of a few months be so multiplied and subdivided, as, that

[1] Latreille *Cours D'Entomologie,* i. 421.　　[2] *Calanus.*

such infinite shoals of beings shall each have a share in the wonderful bequest. But, when we reflect that an Omnipresent Deity is every where mighty in operation, working all in all, and that he guideth all the powers of nature, as the rider guideth the horse upon which he sitteth, to answer the purposes of his providence; [1] we may easily conceive, that under his superintendence the thing may be accomplished, though how it is accomplished, must always remain an unfathomable mystery.

These powers of multiplication are, however, given to these creatures for a wise and beneficent purpose. They themselves afford a supply of food to a variety of creatures—to numerous aquatic insects, even polypes and worms; and to many fishes and birds, by whom their numbers are hourly and greatly diminished. As the stagnant waters likewise, in which they abound, are apt to be dried up in the summer season, many of them probably perish; but, in some, animation may be suspended till the places they inhabit are again filled with water. I have found the little animal described by Dr. Shaw, in the *Linnean Transactions*, as the *Cancer stagnalis* of Linné, in horse-hoof prints, in the spring, then filled with water, but which had been previously quite dry.

---

[1] 1 Cor. xii. 6. Ps. lxviii. 4, 33.

The finny tribes of the world of waters seem more particularly exposed to the invasion of parasitic foes ; as far as they are known there is scarcely a fish that swims that is not infested by more than one of these enemies ; even the mightiest monsters of the ocean, the gigantic whale, the sagacious dolphin, the terrific and all-devouring shark, cannot defend themselves from them. Where they abound they doubtless generate diseases, and are amongst the means employed by a watchful Providence to keep within proper limits the inhabitants of the waters ; and probably there are other benefits which our imperfect knowledge of their history prevents us from duly appreciating, that are conferred, through these animals, upon the oceanic population. Their prevalence upon the predaceous fishes, as was before observed, may tend to diminish their ravages by lessening their activity ; while to those of a milder character, within certain bounds and under certain circumstances, they may be beneficial rather than injurious.

Of this description is the tribe of *Lerneans,* above alluded to as intermediate between the Branchiopod and Pœcilopod Entomostracans ; of which I cannot select a more interesting species to exemplify the adaptation of the structure to the instinct and functions, than one described and figured by Dr. Nordmann, under the appropriate name of *Actheres Percarum*,[1] or *Pest of the Perch*.

This animal, like the Branchiopods, is found in fresh water, where it attaches itself to the common, and another species of the perch genus,[2] and takes its station usually within the mouth, fixing itself, by means of its sucker, in the cellular membrane, so deeply that it cannot disengage itself, or be extracted by external force, without rupturing the so called arms, that are attached to the sucker, and leaving it behind. The animal often fixes itself to the palate, and even to the tongue. The arms[3] take their rise at the base of the cephalothorax—as the part consisting of head and thorax, not separated by a suture, is called—where they are very robust and thick, but they taper towards the other extremity, a single sucker,[4] common to both, being, as it were, hooked to them. These arms are bent nearly into a circle, surrounding the cephalothorax, and the sucker is in front of the head : their substance is cartilaginous, and they repose in the same plane with the head ; whence we may conjecture that the animal, when fixed and engaged in suction, lies close to the part where it has taken its station. When we consider that these predaceous fishes often gorge their prey, swallowing

---

[1] Αχθηρης, *Annoying*.    [2] *Perca fluviatilis* and *P. lucioperca*.
[3] PLATE IX. FIG. 5. *c,c.*    [4] PLATE IX. FIG. 5. *d.*

it entire, we see how necessary it was that our
parasite should be thus fitted to fix itself firmly,
and root itself, as it were, that it may be enabled
to withstand the pressure and violent action of
the bodies that pass over it, for the palate and
tongue of a *Perch* must be a perilous station.
This purpose seems further aided by a quantity
of saliva, usually formed around it.

These pests of the perch are themselves subject to the incursions and annoyance of animals
still more minute than themselves. A small
species of mite[1] makes them its prey, and when
the saliva just mentioned is removed, they are
often found quite covered by a species of Infusory belonging to the genus *Vorticella*.

The next Order, including all the *marine* Entomostracans, will not detain us long. The first
section consists of a single, but very remarkable,
genus, the type of which is the *Monoculus Polyphemus* of Linné.[2] In the West Indies it is
called, by way of eminence, the *King-crab*, and
is found in the seas both of the East and West,
from the equator to the 40th deg. of latitude.
The species are few, and near to each other.
They differ widely both in their characters and
form from every other Crustacean tribe. Like
the *Cirripedes*, they have no distinct head : their

---

[1] *Gamasus scabriculus.*      [2] *Limulus.*—Müll.

crust is divided into two portions, the anterior embracing the posterior, and being terminated, like the Rays, to which they present an analogy, by a long angular tail. They have both compound and simple eyes; the first are situated, one in the middle of each lateral ridge, usually under the spine on the outer side; the second, or simple eyes, are on each side of the intermediate ridge, where it begins: these last are very minute, and not easily discoverable. The under side of the shield, or anterior portion of the crust, is deeply hollowed for the reception of the body, and the cavity is marked out anteriorly by an emarginate ridge, which gives it something the appearance of the hooded serpent. Some of them attain to a large size, the species found near the Molucca Islands being sometimes two feet in length.

The head in them, as in the Arachnidans, seems suppressed, or to merge in the thorax, which also, as in that Class, bears the eyes, the outer pair corresponding with those of certain Crustaceans in which they are sessile, and the inner pair being like those of the Arachnidans, but they have neither the oral organs nor the legs of the Class just named. In fact, these animals seem to stand in much the same position amongst the Entomostracans, that the Cephalopods do amongst the Molluscans, and moreover as giants amongst pigmies. Time will probably throw more light upon these singular works of the Creator.

Their most remarkable organ is their *tail*, which is probably of considerable service to them in their locomotions. It is shaped like a stiletto, and is so extremely sharp at the extremity, that it will easily pierce the flesh, and may perhaps be used by the animal as a weapon, as it is said to be by the Indians; it is so articulated with the posterior piece of the crust as to move with more ease upwards and downwards than laterally. Comparing the small body with the vast volume and levity of the crust which covers and protects it, and considering that the animal, as M. Latreille has remarked, passes the night with its anterior half out of the water, we may conjecture that, by the depression of the tail, it may be elevated in part above the water, and remain stationary. By a slight inclination on either side it probably also helps to steer it, and as it is ciliated at the base, like the natatory legs of a *Dyticus*, it may be of some use in swimming. The legs are all armed with pincers, like those of a crab, from which it seems evident that it is predaceous, and, from their small size, that its prey must consist of minute animals.

The whole of its structure appears calculated to give the king-crab more than usual buoyancy, the reasons of which, when its history is better known, will be more fully understood; and the

Power, Wisdom, and Goodness that every where flash upon us, when we consider animal structures and their adaptation to their habits and instincts, when fully investigated, will be duly appreciated. It is said that this creature, amongst the ancient Japanese, was the symbol of the zodiacal sign *Cancer*.

The animals belonging to the *second* section of the Pœcilipods differ from all the rest, by the manner in which they take their food. They are parasitic upon Cetaceans, fishes, some reptiles, and Crustaceans, whose juices they imbibe by *suction*. They are often fixed to the gills of these animals, but nothing further interesting is known of their history. Some have two long jointed tails, like ephemeræ,[1] and others are distinguished by a remarkable lateral elongation of the thorax.[2] Some fix themselves to their prey by means of suckers, terminating their first pair of legs,[3] which the remainder have not.

The observation of Dr. Von Baer, quoted in a former part of this work,[4] that the lowest grades of the animal kingdom exhibit the leading types of the various organizations it contains, for reasons before alluded to, would almost justify the zoologist in assigning to the Entomostracans a place amongst the Infusories. But the subject of *centres*, in that kingdom, sending forth, as it

---

[1] *Caligus.*    [2] *Nicothoe.*
[3] *Argulus.*    [4] Vol. I. p. 320.

were, rays in different directions, and leading to various forms, requires very deep and minute investigation, and abundant proof, before it will be safe to adopt it as a principle.

## CHAPTER XV.

### *Crustacean Condylopes.*

WE are now arrived at a Class of animals, in which the organs of locomotion assume a new and more perfect form, corresponding in some measure with those of many of the vertebrated animals. The advance, in structure, hitherto, from a mouth surrounded by organs like rays, serving various distinct purposes, and by different means contributing to the nutrition, respiration, and motions of the animal, has been, by certain inarticulate organs, more generally distributed over the body, but still in a radiating order; as for instance, the tentacular suckers of the Stelleridans and Echinidans, which they use in their locomotions, and for prehension, as well as the purposes just named. In the Entomostracans, as we have seen, the legs, though jointed, are very anomalous, assume various forms, and are applied to sundry uses: in the sole instance of the king-crab, they take the articulations of those of the Crustaceans, in which we may trace the general structure of the legs of the other Classes of Condylopes.

But as I shall have occasion, in a subsequent chapter, to give a concentrated account of the gradual developement of the organs of locomotion and prehension, from their first rudiments in the lowest grades of the animal kingdom to their state of perfection in the highest, I shall not here, therefore, enlarge further upon the subject, than by observing, that, in most of the *Decapod* Crustaceans, the anterior legs are become strictly *arms*, terminating in a kind of didactyle hand, consisting of a large joint, incrassated usually at the base, and furnished on its inner side with a smaller moveable one, constituting together a kind of finger and thumb, with which it is enabled to seize firmly and hold strongly any object that its inclinations or fears point out to it. This hand we call the *chela* or claw, or more properly pincers, of the lobster or crab. We find it also in the scorpion and book-crab,[1] which on shore are in some sort analogous to the long-tailed and short-tailed Crustaceans, or lobsters and crabs of the waters. This structure of the hand, in these creatures, is particularly fitted to their wants and situation. A hand like

[1] *Chelifer.*

ours, consisting of a quadruple set of fingers and an opposite thumb, to be of sufficient power for their purposes, must be so disproportioned to their size, as to be an incumbrance rather than a useful instrument of prehension; but as now constructed, it has the requisite strength for the purposes of the animal, without being disproportioned to its size, and inconvenient for its use. Thus we see how nicely every thing is calculated and adjusted by Supreme Wisdom, to the nature and circumstances of every animal form.

But these great claws are by no means universal amongst the Crustaceans. In some the claws are very small, but the loss is often made up to these by an increase as to number, so that if they cannot lay hold of large animals, they can seize, at the same time, several small ones. We have seen that in the king-crab all the legs have these prehensory claws, and they vary in number in many of the smaller Crustaceans, as the shrimp,[1] prawn,[2] pandle,[3] &c. The foreleg of some of these has prehensory claws, that are formed like the mandibles or cheliceres of spiders and the arms of the *Mantis*—whence they are called mantis-crabs. Instead of a forceps, consisting of a finger and thumb, the claw that arms the extremity of the leg is folded down, and received into a channel of the shank,

---

[1] *Crangon vulgaris.*    [2] *Palæmon serratus.*    [3] *Pandalus.*

and kept from dislocation by a tooth, or spine, at the base: this structure may be seen in the shrimp.

There is another circumstance, distinguishing the decapod and stomapod Crustaceans, that is peculiar to them, their eyes are placed upon jointed footstalks, so that when they want to explore and examine what passes around them, they can immediately erect these organs, and so greatly enlarge their sphere of vision, but when they have retired to their retreats in the cavities of the rocks, or to burrows that they have formed, they can place them in repose, in a cavity provided for them by their Creator, in the head.

Any person, who casts an eye over these creatures, will be struck by repeated analogical forms, representing some terrestrial animals of the same Sub-kingdom. Thus a large number of those distinguished by the shortness of their tails, the *crabs*, present, both in their retrogressive and lateral motions and general aspect, an astonishing resemblance to many Arachnidans; some imitating spiders, and others phalangians:[1] and, amongst the long-tailed tribe the lobsters, one[2] very accurately represents a scorpion, and another a mantis.[3]

---

[1] *Macropodia Phalangium.*   [2] *Thalassina Scorpioides.*
[3] *Squilla Mantis.*

We have seen the same tendency in the *Annelidans* to approach or imitate terrestrial forms, as if the marine and aquatic animals were anxious to quit their fluid medium, and to become inhabitants of the dry land. The animal living on shore and in the woods at St. Vincent, taken for a Molluscan by Mr. Guilding,[1] appears almost like a creature that had succeeded in such an attempt.

All these resemblances and approximations show, that the great Creator embraced at one view all the forms to which he intended to give being, and created no individual without furnishing it with organs which give it some relation to others; or so moulding its outward form, as to cause it to represent some others to which it is clear it is not brought near by any characters, common to both, that indicate affinity. What can more evidently and strongly manifest *design*, and that of a mind comprehending simultaneously the whole world of created beings, than thus to concatenate all link to link and wheel within wheel, through all their intricate revolutions and ramifications connecting and connected, and all the while reflecting others of a higher or a lower grade with mimic features? this shows the hand, the art, the wisdom, the power, and the goodness of that un-

[1] VOL. I. p. 347. PLATE VIII. FIG. 1.

fathomable depth and immeasurable heighth of Deity, which comprehends all things and is comprehended by none; and to whom all things owe their being, and their form, and their organs, and their several places and functions.

The general characters of the present class are—

BODY apterous, covered by a calcareous crust, divided into segments. *Legs* jointed, 10—16. *Mouth* composed of a *lip, tongue,* a pair of *mandibles,* often bearing a feeler, and two pairs of *maxillæ,* covered by maxillary *legs. Spinal chord* knotty, terminating anteriorly in a small *brain.* A *heart* and *vessels* for circulation. *Respiration* by gills.

These are divisible into five orders.

1. *Decapods. Gills* situated under the sides of the shell. *Ten* thoracic legs. *Eyes* on a jointed footstalk.

2. *Stomapods. Gills* attached to five pairs of appendages, or spurious legs, under the abdomen. *Eyes* as in the Decapods.

3. *Læmipods.* No abdominal *appendages. Eyes* sessile.

4. *Amphipods.* Head distinct. *Eyes* sessile.

5. *Isopods. Head* distinct. *Eyes* sessile. *Legs* simple, equal.

1. *Decapods.* This order naturally resolves itself into two sections, viz. The *short-tailed*

Decapods or *Crabs*,[1] which have their abdomen folded under the trunk : and the *long-tailed* Decapods or *Lobsters, Cray-fish*, &c.[2] whose abdomen is always extended.

Writers on the Crustaceans usually begin with the short-tailed, and then proceed to the long-tailed Decapods, and this arrangement seems natural, when the transit is to those with sessile eyes, such as the locust-crab ;[3] but yet when we consider how nearly related to the *spiders* the former animals are, and that in the latter, though the head is not formed by a distinct suture dividing it from the thorax, yet its contour is strongly marked out externally by an impression, and internally by a ridge, at least in the lobster and cray-fish,—it seems as if the two tribes should form two parallel lines, and proceed, side by side, towards the Arachnidans and Myriapods.

I shall, however, follow the usual plan, and give now some account of the *crabs*. Of these, none are more remarkable than what have been denominated *land-crabs*, from their usually living on shore, and making for the sea only at certain seasons. Of the most noted species of these I have already given a full account,[4] but I shall here notice some others, having the same habits, that will interest the reader. Aristotle, long ago,

[1] *Brachyuri.*    [2] *Macrouri.*
[3] *Orchesia litterea.*    [4] VOL. I. p. 124.

noticed a crab of this description, found in Phœnicia, under the name of the *Horseman*,[1] which he says runs so fast that it is not easy to overtake it.[2] Olivier found this account true of those he saw on the coast of Syria; and Bosc observed a species[3] in Carolina, which he had some trouble to overtake on horseback and shoot with a pistol. These horsemen crabs are found only in warm climates, where they inhabit sandy spots near the shore, or the mouths of rivers. They make burrows in the sand, to which they retreat when alarmed, and in which they pass the night.

Another kind of land-crab[4] is distinguished by the extraordinary disproportion of its claws; one of them, sometimes the left and sometimes the right, being enormously large, while the other is very small, and often concealed, so that the animal appears single-handed. This formation, however, is not without its use, for, when retired into its burrow, it employs this large claw to stop up the mouth of it, which secures it from intrusion, and this organ is in readiness to seize such animals as form its food and come within its reach. They have the habit of holding up the great one, as if they were beckoning some one; but this doubtless is an attitude of

[1] Ἱππευς. Gr.    [2] *Hist. Anim.* l. iv. c. 2.
[3] *Ocypode Hippeus*, probably *Cancer Cursor*. L.
[4] *Gelasimus vocans*.

defence. These crabs live in moist places, near the shore. They attack, in crowds, any carrion, and dispute the possession of it with the vultures; they do not willingly enter the water, except when they lay and hatch their eggs, and it is conjectured that their young are for some time entirely aquatic. One kind of them,[1] which forms numerous burrows, remaining in them during three or four months in the winter, usually stops them up, so that the animals are obliged to reopen them when the warmth of the vernal sun bids them come forth again from their winter quarters. They are devoured by numerous animals,—otters, bears, birds, tortoises, and other reptiles, all prey upon them, but their multiplication is so excessive, that there seems no sensible diminution of their numbers.

The next tribe of Decapods are the *long-tailed ones*, which do not fold their abdomen under their body. This part is usually furnished at the extremity with several plates, which the animal expands so as to form a fan of five or six leaves; they are easily seen in the common lobster;[2] like the tail of birds, they are useful to the animal in its passage through an element that requires to be moved by organs of a firmer

---

[1] *G. Pugillator.*     [2] *Astacus Gammarus.*

consistence than feathers. The lateral ones in the species just named, having a kind of articulation, so that they can be partially depressed, and push against the plane they are moving upon; they do not, like the crabs, quit the water, and are some of them, as the cray-fish,[1] fresh-water animals.

I shall begin with a tribe which, in some degree, connects the crab with the lobster, these are what are denominated *Hermit-crabs*,[2] whose abdomen being naked, and unprotected by any hard crust, their Creator has given them an instinct, which teaches them to compensate this seeming defect, by getting possession of some univalve shell, suited to their size, which becomes their habitation, and which they carry about with them as if they were its proper inhabitants. These crabs are particularly formed for the habit that distinguishes them. Their naked tail has a tendency to a spiral convolution, fitting them to inhabit spiral shells, which they usually select for their mansion, though, from recent observations, it has been found that any univalve will answer their purpose. Their tail is terminated by an apparatus of moveable and hard pieces,[3] which appear intended to enable the animal to fix itself more firmly in the spire of

---

[1] *Astacus fluviatilis.*  [2] *Pagurus*, PLATE X. FIG. 2.
[3] *Ibid.* 2. *a, a, a.*

the shell. Usually the right hand claw, which is disengaged from the shell, is double the size of the other which is not, and is that which is most employed; but in narrow-mouthed shells, such as the volute, in which Freycinet found one,[1] both claws are disengaged, and are of equal size. The reason of this formation is evident. The fourth and fifth pairs of legs [2] are much smaller and shorter, than the anterior ones, they have, below the claw, a piece resembling a rasp, which appears formed to assist them in moving in the shell, whether they wish to move outwards or inwards, and, on one side, they have a series of egg-bearing appendages.[3] This whole structure proves that they are formed with this particular view of inhabiting the shells of a very different tribe of animals. Some of these hermit-crabs, for there are several species of them, may be called *terrestrial*, while others are *aquatic*. In some of the Indian isles, the shores are covered with them. When the heat is most intense, they seek the shelter of the shrubs, and when the freshness of the evening breathes, they run about by thousands, rolling along their shells in the most grotesque manner, jostling each other, stumbling, and producing a noise by the shock of their encounters, which announces their ap-

[1] *Pagurus clibanarius.*   See PLATE X. FIG. 2.
[2] *Ibid. b b, c c.*              [3] *Ibid. d, d, d, d.*

proach before they appear. When they perceive any danger, they hastily conceal themselves in any ready made holes they meet with, or under the roots, or in the trunks of decayed trees, seldom making for the sea, how near soever they may be. At Guam, a very large species frequents forests more than a mile from the sea; and in Jamaica, another species, called there the soldier,[1] has been found in great quantities on elevated ground, more than four leagues from it.

The common species[2] is aquatic, and usually inhabits the whelk; it is stated annually to leave its shell, at the time of its moult, and after this great crisis is over, to seek another suited to its increased magnitude. Aristotle, Belon, and others affirm that these animals quit their shell to seek their prey, and that when danger threatens them, they retreat to it backwards, but observations have not been made by modern authors which confirm this statement. Their sexual intercourse, however, could not take place without their first leaving their mansion.

Why our, so called, hermits are gifted with this singular instinct, is not easy to conjecture. Many other creatures make use of houses that they had no hand in erecting, as the bees, the cuckoo, and sometimes the bear, &c.; but I do not recollect any that, as it were, clothe themselves

---

[1] *Pagurus Diogenes.*    [2] *P. Bernhardus.*

with the cast garments of other animals. Providence, besides the defence of their otherwise unprotected bodies, has no doubt some object of importance in view in giving them this instinct. Perhaps they may accelerate the decomposition of the shells they inhabit, and cause them sooner to give way to the action of the atmosphere; and as all exuviæ may be termed nuisances and deformities, giving to these deserted mansions an appearance of renewed life and locomotion, removes them in some sort from the catalogue of blemishes. By this physical hypocrisy, of assuming the aspect of a different animal, which is known as not having powerful means of destruction, these creatures may deceive the unwary, and make them their prey, which if they wore the livery of their own tribe, would be on their guard and escape them.

Next to the Hermit-crabs, or rather Hermit-lobsters,[1] comes a very interesting genus, which might be denominated *Tree-lobsters*, from the singular circumstance of their quitting the sea, like the Climbing-perch,[2] and in the night ascending the cocoa-nut, and other palm-trees, for the sake of their fruit. The species which manifests this remarkable instinct is gigantic, and must exhibit a striking spectacle when engaged in ascending the stem of a cocoa-tree; but Mr. Cummings ob-

---

[1] *Birgus Latro.* PLATE X. FIG. 1.    [2] VOL. I. p. 123.

served its proceedings in the Polynesian Islands, where he saw it ascending the palm-trees and devouring their fruit. I have, in a former chapter,[1] stated that the *Climbing perch* ascends the fan-palm in pursuit of certain Crustaceans, perhaps related to the *Birgus*, which frequent it. Freycinet observed these crabs, in the Marian Islands, and says that their claws have wonderful strength, for when the animal has seized a stick, an infant may be suspended from them. They are very fond of the fruit of the cocoa-palm, and may be fed with it for months without suffering from want of water. Whether, like the land-crab, they have a reservoir capable of containing a sufficient quantity of that fluid to keep the gills moist, has not been ascertained: probably they have.

Amongst the larger species of the long-tailed Section, there is one of a most ferocious aspect, having its head, the base of its long antennæ, and its thorax, beset with sharp spines. This is called in the London market the *Thorny lobster*,[2] and is stated sometimes to be nearly a yard in length: it is also called the *Cray-fish*, and by the French, who esteem it highly, the *Langouste:* it is, however, far inferior to the common lobster, from which it is distinguished by having no pincers, its legs terminating in a strong simple

[1] Vol. 1. p. 126.

[2] *Palinurus vulgaris*, Leach. *Malacostr. Podophth. t.* xxx.

claw, set with bunches of bristles, a circumstance indicating a different mode of taking its prey. From the amplitude of their fan-like tail, and from their natatory plates, these lobsters seem formed for rapid motion in the water.

The next species that I shall mention is of much more importance to us, and has been celebrated by epicures from ancient times. Instead of unarmed hands and legs, the *Lobster*,[1] as every one knows, has the *former* armed, often with an enormous pair of claws, which must be of vast power, and, besides, the two anterior pairs of their *legs* are furnished with small pincers. It is observable that the moveable finger of the claw of the hands is on their inner side, while, in these two pairs of legs, that on the outside is moveable. Aristotle's *Carabus*[2] is generally referred to the thorny lobster; but in one place he expressly mentions its using its pincers to catch and carry its food to its mouth, which could not apply to that animal, though it agrees well with the common lobster; yet in another place, under the same name, he appears to mean the other.[3] It is not known exactly to what use these smaller pincers are applied; it must be observed, however, that if the legs are regarded as naturally pointing towards the head, as in Dr. Leach's figure of *Nephrops*, the moveable

---

[1] *Astacus Gammarus.*

[2] Gr. καραβος, *Hist. Anim.* l. viii. c. 2.    [3] *Ibid.* l. ii. c. 2.

thumb in all is on the same side. The antennæ in this genus are about the length of the body. The pincers of the hand are very powerful and tubercular; they are used by these animals both to seize their prey and for self defence, and they contain very powerful muscles. When in the water the lobster seizes anything presented to it, and holds it so strongly that it is impossible to extricate it without breaking the claw.

All Crustaceans cast their crust annually. At first it seems wonderful how this can be accomplished. With insects, in whom it takes place only in the larves, and whose form and substance are usually adapted to it, a longitudinal fissure of the skin of a soft caterpillar, or grub, when the animal grows too big for it, we can conceive to be no difficult task: but with animals covered with a hard crust, and in whom not only the covering of the head, trunk, and abdomen is to be cast, but also that of the legs and other organs, it seems an operation infinitely more arduous. But HE who gave them this defence, instructs them also how to rid themselves of it when it grows too strait for them, and has moulded their structure accordingly.

These animals are not, like most insects, limited to an existence, terminated within the period of one revolution of the earth round the sun, but sometimes witness several; and some are said even to live *twenty* years, and keep growing during the greater part of their life. But this would be impossible, since it is incapable of extension, unless they could give room for the expansion of their body, by occasionally rejecting the case which incloses it. At a certain time of the year, about the end of the spring, when food is plentiful, they begin to feel themselves ill at ease: they then probably seek the clefts of the rocks, and other close places, in which they can undergo, in concealment and security, a change which exposes them, in a defenceless state, to danger.

· But we should have known nothing of the manner in which this great work is effected, had not the illustrious French naturalist, Reaumur, adopted methods which enabled him to ascertain their mode of proceeding. In the spring, in boxes pierced with holes, which he placed both in the river, and in an apartment, he put the fresh-water cray-fish,[1] of the same genus with the lobster. He observed that when one of these was about to cast its crust, it rubbed its feet one against the other, and gave itself violent contortions. After these preparatory movements, it swelled out its body more than usual, and the first segment of its abdomen appeared more than commonly distant from the thorax. The membrane that united them now burst, and its new body appeared. After rest-

[1] *Astacus fluviatilis.*

ing for some time, it recommenced agitating
its legs and other parts, swelling to the utmost
the parts covered by the thorax, which was
thus elevated and separated from the base of
the legs; the membrane which united it to the
underside of the body burst asunder, and it only
remained attached towards the mouth. In a few
minutes, from this time, the animal was entirely
stripped except the legs. First the margin of
the thorax was seen to separate from the first
pair of legs; at that instant, drawing back its
head, after reiterated efforts, it disengaged its
eyes from their cases, and all the other organs
of the anterior part of the head; it next uncased
one of its fore legs, or all or part of the legs of
one side, which operation is so difficult that
young ones sometimes die under it. When the
legs are disengaged, the animal casts off its
thorax, extends its tail briskly, and pushes off
its covering and that of its parts. After this
last action, which requires the utmost exertion
of its remaining vigour, it sinks into a state
of great weakness. Its limbs are so soft that
they bend like a piece of wet paper; but if
the back is felt, its flesh appears unexpectedly
firm, a circumstance arising, perhaps, from the
convulsive state of the muscles. When the
thorax is once disengaged, and the animal has
begun to extricate its legs, nothing can stop its
progress. Reaumur often took them out of the

water with the intention of preserving them half
uncased, but they finished, in spite of him, their
moult in his hands. Upon examining the
exuviæ of these animals, we find no external
part wanting; every hair is a case which covers
another hair. The lower articulations of the
legs are divided longitudinally at a suture which
separates during the operation, but which is not
visible in the living animal.

When we consider this apparently arduous
and complex operation, we see the most evident proofs of *design*, and that the Creator
has so put together the different parts of the
animal's structure, that there is no occasion to
divide the crust itself in order to liberate it.
Instead of a solid tube, he has inclosed the leg
in joints that are furnished with the means of
dividing longitudinally, upon sufficient expansion of the included limb, and so opening a way
for its liberation. In the whole body all the
segments and parts are so united by a membrane which can yield to the expansive efforts of
the animal, that the entire liberation of it from
the armour that encases it, is accomplished with
infinitely more ease than we should expect, even
after a careful investigation of it. Besides
membranous ligaments, so arranged by the Wisdom of the Creator as to yield to the efforts of
these creatures to liberate themselves from their
too strait garment, he has also furnished them,

as Reaumur remarks, with a slimy secretion, which moistens the interval between the old and new shell, and facilitates their separation.

The time requisite for hardening the newly acquired crust, according to its previous state, is from one to three days. Those animals that are ready to moult have always two stony substances called crabs'-eyes, placed in the stomach, which, from the experiments of Reaumur and others, appear destined to furnish the matter, or a portion of it, of which the shell is formed, for if the animal is opened the day after its moult, when the shell is only half hardened, these substances are found only half diminished, and if opened later they are proportionably smaller. Thus has Creative Wisdom provided means for the prompt consolidation of the crust of these creatures, so that it is soon rescued from the dangers to which, in its naked state, it is exposed. Reaumur measured several cray-fish, before, and after their moult, and found that their augmentation amounted to about one-fifth, this amount probably decreases as they approach nearer to their adult state. From a chemical analysis of the crust of the lobster it has been ascertained that it consists of gelatine united to calcareous earth ; it differs from the shells of Molluscans in having a much greater proportion of gelatine, whereas in the latter the calcareous earth greatly predominates.

It is asserted that *birds*, and other animals in tropical countries, have *two* moults within the year, after the two rainy seasons are passed, and two broods ; whether this is the case with Crustaceans has not been ascertained. Most other Condylopes do not survive the laying of their eggs, but the Crustaceans are evidently exempted from this law, and emulate the higher animals in the duration of their existence.

It may be observed that the moult of Crustaceans differs in one respect from that of birds, which only change their feathers, and that of quadrupeds who only change their fur, since they disengage themselves from their whole external skin with all its appendages, whether of fur, or any other substance. Their moult resembles rather that of trees, whose outer skin, under the form of bark, peels off annually, and is succeeded by another formed under it, as is particularly evident in the birch, plane, &c.

It is to the researches of the same learned, and patient, and penetrating experimenter and naturalist that we are indebted for what knowledge we possess of the means employed by nature for the reproduction of the mutilated organs of Crustaceans. Having cut off the legs of some crabs and lobsters, and placed them in covered boats, communicating with the water, and destined to keep fishes or Crustaceans alive, at the end of some months, he saw that the mutilated legs had been replaced by new ones, perfectly resembling the old, and almost as large. The time necessary for this reproduction was not fixed, but depended upon the warmth of the season, and the supply of food furnished to the animal, and likewise upon the part in which the mutilation took place. The point of union of the second and third joints, is the part of the leg where a fracture is most easily made, and the reproduction is most rapid. At this point there are many sutures which appear distinct from articulations; it is in these sutures, particularly the intermediate one, that the separation usually occurs, and many Crustaceans, if they are wounded in some other part of their leg, cast the remainder off at this suture to facilitate the reparation of their loss. So much only is reproduced in each leg as is necessary to render it again complete.

When a leg is mutilated in the summer, if examined a day or two after the experiment, the first circumstance observable is a kind of covering membrane of a reddish hue; in five or six days more this membrane becomes convex; next it is protruded into a conical shape, and keeps gradually lengthening as the germinating leg is developed; at last the membrane is ruptured and the leg appears, at first soft, but in a few days it becomes as hard as the old one; it now wants only size and length, and these it acquires in time; for at every moult it augments in a more rapid proportion than the legs that have their proper size. The antennæ, maxillæ, &c., are reproduced in the same manner, but if the tail is mutilated, it is never reproduced, and the animal dies. When attacked, Crustaceans, as well as some of their analogues, the grasshoppers, often cast their legs as it were voluntarily.

When we reflect on this history, we cannot help admiring and adoring the goodness of the Creator, and his care over the creatures he has made, in giving to these animals, which, both from the multiplicity and exposure of their legs, and other organs, and their numerous enemies, are particularly liable to mutilations, a power that enables them, in a short period, to pursue the course directed by instinct, with undiminished or little diminished powers.

The *Stomapods*, or mouth-legged Crustaceans, so named because the maxillary legs do not differ materially from the thoracic ones, form the *second* Order of the Class, and the species belonging to it, on account of their general resemblance to the orthopterous tribe forming Linné's genus *Mantis*, are called *Sea-Mantises*. One of them,[1] in its anterior legs, accurately

---

[1] *Squilla Mantis.*

represents that genus. But the most remarkable animals belonging to the Order are the *Phyllosomes*[1] of Dr. Leach, which in some respects are analogues of the *Spectres*,[2] not having the raptorious fore leg of the Squillæ, but their thorax, which consists of two segments, the first very much dilated, approaches nearer to that of *Mantis strumaria*.[3] It has been taken in several tropical seas, and when living, it is said to be as transparent as crystal, except its eyes, which are sky-blue.

The subsequent Orders of the Crustaceans, called by the general name of *Malacostracans*, are distinguished from the preceding by having sessile eyes, imbedded in the substance of the head, and though they contain many singular creatures, we know little of their habits and history.

Many of the animals belonging to Latreille's *Læmodipods*, or throat-footed Crustaceans, which begin the sessile-eyed tribes, have very slender bodies, and their legs are separated by a considerable interval, like those of geometric larves or loopers amongst insects, whose motions they also imitate. One remarkable creature is included in this Order, which is parasitic upon the whale,[4] and by its hooked claws is enabled to

---

[1] PLATE X. FIG. 3. *P. brevicorne?*

[2] *Phasma.*          [3] Stoll. *Spectr. t.* xl. *f.* 42.

[4] *Cyamus Ceti.*

maintain its station amidst the fluctuations of the waves. This animal, like the king-crab, has both compound and simple eyes.

Next to these succeed the Order of *Amphipods*, including a number of genera, consisting usually of minute animals; many of them, like the grasshoppers, and several other insects, are gifted by their Maker with the faculty of leaping. When one meets with a heap of sea-weeds upon the beach, recently left by the tide, if we turn it over we shall often see under it myriads of little animals belonging to this Order, jumping about in all directions, which are thus enabled, either to find shelter under another mass of moist sea-weed, or perhaps to reach their native waves in safety. Whether these Crustaceans, like their analogues on shore, feed on vegetable substances, has not been ascertained; they are generally found as above stated; and there may be *herbivorous* species amongst the Crustaceans, as well as in almost every other class of animals.

The last Crustacean Order is called by Latreille, *Isopods*, from their legs being usually of the same length; though a large proportion of these are *aquatic* animals, yet the Order terminates in those that are *terrestrial*. Several of the former are furnished with one or more pair of didactyle legs, but the terrestrial ones never have these prehensory organs.

Amongst the Crustaceans, Latreille has included the *Trilobites,* a remarkable tribe of animals, at present found only in a fossil state, and like the chitons, certain wood-lice,[1] and the armadillo,[2] rolling themselves up in a ball. They may form part of a branch connecting the Crustaceans and Molluscans, but I leave the discussion of this point to abler hands.

Thus have we at length arrived at animals, the majority of which are *terrestrial,* at least in their perfect state, for many terrestrial Condylopes have aquatic larves and pupes, but few, or none, I believe, inhabit salt water, except perhaps some species of bugs.[3]

The great Crustacean host, of which probably we do not know half the species, is certainly a most valuable gift to mankind, as well as to the various inhabitants or frequenters of the waters, especially of the ocean, varying as they do in size, from the great thorny lobster to the minute tribes of Entomostracans; they probably become the prey of many sea animals, besides the Cephalopods, which are stated to make such havoc among them.[4] When we further consider their powers of infinite multiplication, we see that however great the consumption of them, there appears no diminution of their numbers, so that one kind of animals, by the will of Him who

---

[1] *Armadillo vulgaris.*      [2] *Dasypus.*

[3] *Salda Zostaræ.* F. &c.     [4] See above, VOL. I. p. 314.

created all things, and who gave a law to each species, which regulated their numbers, and the momentum of their action, doing or suffering, is made to compensate for another, and the law of preservation to act as an equipoise to the law of destruction.

When we look, however, at these animals, especially the larger kinds, and survey their offensive organs and weapons, and the coat of mail that defends them, we feel convinced that they also are employed to keep down the numbers of other inhabitants of the ocean, more especially as the great body of them are evidently predaceous : and this, on such a survey, seems to us their primary function. God numbers and weighs them both with those they destroy and those that destroy them; his bridle is in their mouth, and they go as far as he permits them : and when he gives the word—Peace, be still—the mutual conflict relaxes, or, in some parts, is intermitted, till the general welfare calls for its revival.

It may be observed with regard to this constant scene of destruction, this never universally intermitted war of one part of the creation upon another, that the sacrifice of a part maintains the health and life of the whole ; the great doctrine of *vicarious suffering* forms an article of physical science ; and we discover, standing even upon this basis, that the sufferings and death of one being may be, in the Divine counsels, and consistently with what we know of the general operations of Providence, the cause and instrument of the spiritual life and final salvation of infinite hosts of others. Thus does the animal kingdom, in some sort, PREACH THE GOSPEL OF CHRIST.

## CHAPTER XVI.

*Functions and Instincts. Myriapod Condylopes.*

THERE are two Classes of Condylopes, extremely dissimilar in their external form and the number of their legs, and yet in some respects related to each other, at each of which we may be said now to have arrived; both are almost exclusively terrestrial, and both remarkable for their ferocious aspect; the one the analogue of the *crab*, and the other apparently related to the Isopod Crustaceans, the *oniscus* and *armadillo*. It will be easily seen that I am speaking of the *Arachnidans* and *Myriapods*.

Regarding, therefore, the long-tailed Decapod Crustaceans as leading, by the Order of Isopods which we last considered, towards the *Myriapods,* and the short-tailed ones or crabs, as tending towards the *Arachnidans.* I shall give a brief account of the former of these Classes in the present chapter, and I am the more induced to assign them precedency because of their evident connection with certain *Annelidans*, which indeed Aristotle, and other ancient Naturalists, thought was so close, that they considered them as belonging to the same genus,[1] and it is worthy of remark that, in the Class just named, the representatives, if they may be so called, of the Myriapods, are, like them, divided into two tribes, one with a *cylindrical* and the other with a *flat* body.[2]

The Myriapods exhibit the following general characters.

ANIMAL undergoing a metamorphosis by acquiring in its progress from the egg to the adult state several additional segments and legs. *Body* without wings, divided into numerous pedigerous segments, with no distinction of trunk and abdomen. *Head* with a pair of antennæ; two compound eyes; a pair of mandibles; under-lip connate with the maxillæ.

The class naturally divides itself into two *Orders*, distinguished both by their form and habits.

1. *Chilognathans.*[3]  BODY generally cylindri-

---

[1] Aristot. *Hist. Animal.* l. ii. c. 14.  Plin. *Hist. Nat.* l. ix. c. 43.

[2] See VOL. I. p. 347, and PLATE VIII. FIG. 1. 4.

[3] *Chilognatha*, so called because their *lip* is formed of the jaws, from Gr. χειλος, a lip, and γναθος, a jaw.

cal; segments half membranaceous and half crustaceous, each half bearing a pair of legs; *antennæ* seven-jointed, filiform, often a little thicker towards the end. These are called Millipedes. *Julus L.*

2. *Chilopodans.*[1] *Body* depressed; segments covered by a coriaceous plate, bearing each only a single pair of legs; *antennæ* of fourteen or more joints, setaceous. These are called Centipedes. *Scolopendra L.*

1. Very little is known with respect to the habits and instincts of the animals belonging to either of these Orders, except that they frequent close and dark places, being usually found under stones, under bark, in moss, and the like.

Latreille names the three families into which he divides the *first* of them, *Onisciform, Anguiform,* and *Penicillate;* one[2] resembles a woodlouse, like the mammalian armadillo, the trilobites, and chitons, when alarmed, rolls itself up into a spherical ball; besides the ordinary dorsal and ventral segments, these have, on each side underneath, between the lateral margin and the legs, a series of rounded plates, which Latreille conjectures may be related to the organs of respiration, which seems to give them some further affinity to the Trilobites.

---

[1] *Chilopoda,* so called because their *lip* is formed of the *foot,* from Gr. χειλος, a lip, and πυς, a foot.

[2] *Glomeris.*

They are found mostly under stones, and creep out before rain.

Another,[1] in its cylindrical body, gliding motion, and coiling itself up spirally, presents a striking resemblance to a snake. Some species[2] emit, through pores, that have been mistaken for spiracles, a strong and rather unpleasant odour.

The *penicillate* family, of which only a single species is known,[3] is remarkable for several pencils or tufts of long and short scales, which distinguish the sides of the body. These are found principally under the bark of trees.

The myriapods belonging to this order De Geer describes as very harmless animals. They appear to feed upon decaying vegetable or animal matter. The author just named thinks that the common *Julus*,[4] or Gallyworm, feeds upon earth; one that he kept devoured a considerable portion of the pupe of a fly; other species are stated to eat strawberries and endive; and Frisch fed one, that he kept a long time, upon sugar.

2. The *Chilopodans* or Centipedes, which constitute the *second* order, Latreille divides into two families, which he denominates *Inæquipedes* and *Æquipedes*. The *Inæquipedes*, so called be-

---

[1] *Julus*, &c.          [2] *J. fœtidissimus.*
[3] *Pollyxenus lagurus.*    [4] *J. terrestris.*

cause the six last pairs of legs are suddenly longer than the rest, belong, as at present known, to a single genus,[1] which being less depressed than the other Centipedes, seems to connect the two Orders. They are not found in England, but in France they are stated to frequent houses and outbuildings, where they conceal themselves during the day, between the beams and joists, and sometimes under stones ; but when night comes they may be seen running upon the walls, with great velocity coursing their prey, which consists of insects, woodlice, and other minute creatures; these they puncture with their oral fangs, and the venom they instill acts very quickly, thus enabling them easily to secure their victim.

The *Æquipedes*, so called because all their legs, except the last pair, are nearly equal in length, are sub-divided into several genera, the most remarkable of which is distinguished by the ancient name of *Scolopendra*. Some species of this genus grow to an enormous size ; a specimen of the giant centipede[1] in the British Museum is more than a foot long. The arms of the animals of the present Order are more tremendous than those of the Millipedes, for their second pair of legs terminates in a strong claw,[2] which

---

[1] *Cermatia.* Illig. Leach.   *Scutigera.* Lam. Latr.
[2] *Sc. Gigas.*        [3] *Introd. to Ent. t.* vii. *f.* 13. *ä.*

is pierced at the apex for the emission of poison ; in this family the first or hip-joints of these legs are united and dilated so as to form a lip.[1]   In warm climates, the centipedes are said to be very venomous.

As the anguiform *Chilognathans* represent the living and moving serpent, so the family I am now considering, the equipede *Chilopodans*, may be regarded as representing the skeleton of a dead one.   The head, with its poison-fangs, the depressed body, formed of segments representing vertebral joints, and the legs curving inwards, and resembling ribs, all concur to excite the above idea in the mind of the beholder.

Like the last family, these also frequent close places, and sometimes creep into beds; they devour insects, and similar small animals, which Latreille found the puncture of their envenomed fangs arrested, and killed instantaneously; and it is sometimes attended with serious inconveniences to man himself.   One species,[2] in some parts of the West Indies, goes by the name of the *Mischievous;* [3] and the pain caused by the bite of the Giant Centipede, though it is never mortal, is greater than that produced by the sting of the scorpion.

Some centipedes emit a phosphoric light; of this description is one distinguished by the name

[1] *Introd. to Ent. Pl.* vii. *f.* 11. *d, b.*
[2] *Scolopendra morsitans.*          [3] *Malfaisante.*

of the *phosphoric*,[1] which is stated by Linné to have fallen from the air upon Captain Ekeberg's vessel in the Indian Ocean, a hundred miles from land. But the light-giving centipede best known is the *electric*,[2] which is remarkable for emitting a vivid phosphoric light in the dark; this is produced by a viscid secretion, which, as I have observed, when adhering to the fingers, gives light independently of the animal. This species also frequents beds. Its object in this may, perhaps, be to search for bugs and other insects that annoy our species during repose.

The function which the Creator has devolved upon the Myriapods of the first Order, seems to be that of removing *putrescent* vegetable and animal matter from the spots that they frequent; and that of the second to keep within due limits the minor inhabitants, especially the insect, of the dark places of the earth. Viewed in this light, however disgusting they may seem to us in their general aspect, we may regard them as beneficial, and as contributing their efforts to maintain in order and beauty the globe we inhabit.

It is worthy of remark that the great Hebrew Legislator, amongst the unclean animals which it was unlawful for the Israelites to eat or to

---

[1] *S. phosphorea.*    [2] *Geophilus electricus.*

touch, enumerates those which *multiply feet.*[1] In the common version it is translated, *Hath more feet;* but the marginal reading is nearest to the Hebrew,[2] and seems to allude to a circumstance upon which I shall hereafter enlarge, namely, that these animals increase the number of their legs with their growth. As a subject intimately connected with Zoology in general, and leading to a very profitable study of the animal kingdom in a moral point of view, it will not be foreign to the object of the present treatise if I add here a few remarks upon the distinction of animals into clean and unclean, observable in many parts of Holy Writ. This distinction was originally to indicate those which might or might not be offered up in sacrifice, and, afterwards, when animal food was permitted, to signify to the Jews those that might and those that might not be eaten. When Noah was commanded, *Of every* clean *beast thou shalt take to thee by sevens, the male and his female; and of beasts that are* not clean, *by two, the male and his female.*[3]—it is evident that the distinction was familiar to the Patriarch. The *unclean* animals, with respect to their habits and food, belonged to two great classes, namely *Zoophagous* animals, or those which attack and devour *living* animals; and *Necrophagous* animals, or those which devour

---

[1] *Levit.* xi. 42.      [2] מרבה רגלים      [3] *Genes.* vii. 2.

*dead* ones, or any other putrescent substances. Of the first description are the *canine* [1] and *feline* [2] tribes amongst *quadrupeds*; the *eagles* [3] and *hawks* [4] amongst *birds*; the *crocodiles* [5] and *serpents* [6] amongst *reptiles*; the *sharks* [7] and *pikes* [8] amongst *fishes*; the *tiger-beetles* [9] and *ground-beetles* [10] amongst *insects*; and to name no more, the *centipedes* in the class we are treating of.

With regard to the *necrophagous* tribe, I do not recollect any *mammalians* that are exclusively of that description, for the *hyæna* [11] and *glutton* [12] are ferocious, and eagerly pursue their prey, they will, however, devour any *carcasses* they meet with, and even disinter them when buried; but the *vulture* amongst the *birds* will not attack the *living* when he can gorge himself with the *dead*; the *carrion crow* belongs also to this tribe; amongst *insects*, the *burying*, [13] *carrion*, [14] and *dissecting beetles*, [15] the *flesh-fly*, and many other *two-winged* flies, feed upon *putrescent flesh*; and numberless others satiate themselves with all unclean and putrid substances, whether animal or vegetable. In the present class, the *millipedes* belong to the necrophagous tribe.

[1] *Canis*  [2] *Felis.*  [3] *Aquila.*

[4] *Falco.*  [5] *Sauria.*  [6] *Ophidia.*

[7] *Squalus.*  [8] *Esox.*  [9] *Cicindela.*

[10] *Carabus, Harpalus,* &c.  [11] *Canis Hyæna, L.*

[12] *Necrophorus.*  [13] *Silpha.*  [14] *Dermestes.*

[15] *Sarcophaga carnaria.*

A third description of animals, appearing to be intermediate between the clean and unclean, and partaking of the characters of both, was added to the list—for instance, those that are *ruminant* and do *not divide the hoof*, as the *camel*, which, though it has separate toes, they are included in an undivided skin; and those that *divide the hoof*, but are *not* ruminant, as the *swine*.

It appears clear from St. Peter's vision, recorded in the Acts of the Apostles,[1] that these unclean animals were symbolical, and in that particular case represented the Gentile world, with whom it was not lawful for the Jews to eat or associate,[2] doubtless, lest they should be corrupted in their morals or faith, and seduced into Idolatry, and its natural consequences, with regard to morality, by them. In other passages of Scripture, unclean animals are employed to symbolize evil and unclean *spirits* as well as *men*, as the serpent, the dragon, or crocodile,[3] the lion,[4] and the scorpion.[5]

By way of corollary to the present short chapter, I shall devote a few pages to a very interesting subject, intimately connected with the animals whose history and habits I have just described, and which marks out the plan upon which the wisdom, power, and goodness of the

---

[1] *Acts*, x. 10—15.    [2] *Ibid.* ver. 28.    [3] *Revel.* xx. 2.
[4] 1 *Pet.* v. 8.    [5] *Luke*, x. 19.

Creator have been manifested in animal structures. I allude to what has been named the *conversion* of organs, by which term is meant, not only in particular instances, multiplying the functions of any given organ, as, for instance, when the *tail* of an animal is employed like a *hand,* to take hold of the branch of a tree, and so assist in locomotion, as in the chameleon, and certain monkeys;[1] and the tongue is also made to subserve to prehension, as in the case of the giraffe; but likewise when the organ is converted from one use to another, as when the anterior leg is taken from locomotion, and given to prehension, as the human hand; or as when all the ordinary organs of locomotion in one tribe are in another converted into oral organs, either to assist in mastication, or to discharge the office of a lip, as in the Crustaceans and centipedes. In the investigation of this curious and interesting subject, the class of Myriapods affords an example, if I may so speak, of the gradual conversion of locomotive organs into auxiliary oral ones. Something of this kind I have before stated,[2] is discoverable in certain Annelidans, either related to those animals or their analogues.

In the *Introduction to Entomology* it is observed, with respect to the larves of many *Hexapod* Condylopes, that their progress towards

---

[1] *Ateles.*   [2] See above, Vol. I. p. 346.

what is called their perfect state, is by *losing* their spurious *legs* or *prolegs*, and by *acquiring* organs of *flight;* whereas in the *Myriapods*, the reverse of this takes place ; instead of losing legs and shortening their body, some of them when first hatched, have only *six* legs, representing the six legs of Hexapods, and all in their progress to their adult state acquire a large number of what may be denominated spurious legs, which support many additional segments.

As the *Chilognathans*, in their young state, come nearest to the insect or hexapod tribes, I shall begin by stating the changes they undergo. In the most common species,[1] according to De Geer's description and figure, the animal is divided into three principal parts, as in Hexapods ; first, there is a *head* with antennæ, and the usual *oral* organs, though a little aberrant in their structure ; next, there is a *trunk*, consisting of three segments, each bearing a pair of legs ; and lastly, there is an *abdomen*, divided into five segments, without legs.[2] With regard to their oral organs, they correspond with those of Hexapods, both in number and kind, for in the mouth, above is a representative of the upper-lip ; below this is a pair of mandibles or upper-jaws ; next follows a lower-lip, consisting of three pieces united together, the two lateral ones analogous

---

[1] *Julus terrestris.*    [2] De Geer, vii. 583. *t.* xxxvi. *f.* 20, 21.

both to the lower-jaws of Hexapods, and the first pair of maxillæ of Crustaceans ; and the intermediate one, resolvable into two pieces, representing the lip of the former and the second pair of maxillæ, according to Savigny, of the latter, from his figures,[1] the maxillary and labial feelers appear to have their representatives ; yet though he has figured he does not notice them as feelers.[2]

The six original or natural legs of the Iulus are its first organs of locomotion, which when the animal is arrived at its complete develope- ment, as to number of legs and segments,—are said still to maintain their original function, although probably diminished in energy ; the two first pairs are, however, as it were, applied to the mouth, the segments that bear them being very short. The sciatic joint or hip[3] of the first pair forms a single piece ; those of the second are also united and more elevated ; but those of the third are distinct: so that in this Order of the Myriapods we see the first tendency towards employing what in Hexapods wear the form and perform the functions of *legs* as auxiliaries of the *mouth*, and of the locomotive function being devolved upon organs which have no represen-

---

[1] *Anim. sans Vertébr.* Mem. ii. *t.* i. *f.* 1. o. 2. o.

[2] He says that the pieces forming the labium are *Dénuées des palpes. Ibid.* p. 44.

[3] *Coxæ.*

tative in Hexapods, except in their incipient state.

To proceed next to the *Chilopodans*—it has not yet been ascertained what changes they undergo in the progress of their growth, save that the number of legs and segments increases till they have arrived at their full size,[1] nor is it known how many they have when first hatched, but, from their structure, it seems evident that the analogues of the two first pair of legs of the Chilognathans, can never be employed in locomotion; and further, that not only is their first or hip-joint united with its fellow, so as to form a kind of auxiliary lip, but the other articulations are converted into prehensory organs, instead of a locomotive one, in the first pair armed at the end with a minute forceps, and in the second with a fang resembling the tooth of a serpent, having a pore at the extremity for the emission of poison, connected with an *Ioterium* or poison bag.

Here then, in these two Orders of the Myriapods, we have a regular *conversion* of organs: those that in the Millipedes are used for locomotion, in the Centipedes, exchange that function for that of prehension, both agreeing in being auxiliary, at their base, to mastication, but the latter with a greater momentum.

The reason of this change in the functions of

[1] De Geer, vii. 562.

these organs we shall readily see when we consider the habits and food of these respective
Orders. The Chilognathans deriving in general their nutriment from *putrescent* substances
whether animal or vegetable, have no resistance
to overcome, and therefore require not the aid of
additional prehensory organs to enable them to
execute their offices; while the Chilopodans,
having to contend with *living* animals, must put
them *Hors de combat*, either by killing them, or
deadening their efforts, before they can devour
them. In this last Order we find that though
the two first pairs of legs have a new office, the
third pair are still used for locomotion.

From the oral organs and their auxiliaries of
the Myriapods to those of the *Crustaceans,* the
interval is not very wide ; and amongst the latter
the *Isopods,* especially the terrestrial ones, as
might be expected, approach the nearest to
them. De Geer observes that the common
wood-louse,[1] which in its adult state has fourteen
legs; when it first leaves the egg, has only six
pairs and six segments ;[2] thus doubling the
number of the Hexapods and *Julus;* and in this
animal and its relation, *Ligia,* the thoracic legs
are all used in locomotion ; but when we examine the *aquatic,* especially the *marine,* genera
of this Order, as *Idotea, Stenosoma,* &c., we find

[1] *Oniscus Asellus.*  [2] vii. 551.

that the first pair of thoracic legs is taken from that function, and made auxiliary to the organs of the mouth.

Leaving the Isopods, if we go to the *Decapods*, amongst those with a long tail,[1] which from their cylindrical form and other circumstances, are nearer to the Chilognathan Myriapods than to the Chilopodan, taking the lobster for our type, we find the organs analogous to the six legs of Hexapods, exhibiting a new character: for from the outer side of their basal joint issues an organ which is peculiar to these legs. The organ I allude to is called, by M. Savigny, a *flagrum* or whip; and, by M. Latreille, a *flagelliform palpus* or feeler; it usually consists of two parts, an elongated exarticulate base, representing the *handle* of the whip; and an annulated or jointed part generally forming an angle with it, representing the *lash:* the mandibles also have feelers of the usual structure. The organs above alluded to, shew that all the representatives of the legs of Hexapods in the lobster, are converted to a new function—whether precisely analogous to that of feelers is not clear.

In the lobster the basal joints of the first pair of maxillary legs are dilated, and the whole organ may be regarded as maxilliform; but in the second it is palpiform, and in the third it resumes the joints and appearance of a crus-

[1] *Macrouri.*

taceous leg, and is densely ciliated, which seems to indicate that it is used in swimming.

In the common crab,[1] amongst the short-tail Decapods,[2] the legs in question seem all taken from locomotion, and the second pair does not differ from those of the lobster; but the last, though consisting of the same number of joints, is very different, the two intermediate joints being dilated, and the two legs together forming as it were a pair of folding-doors, which close the mouth externally, the three last joints resembling those of the legs. These animals, therefore, in some sort, the flatness of their body and this double auxiliary lip considered, present the same analogy to the *Chilopodan* Myriapods, that the lobster does to the *Chilognathan*. In both we see, by their feelers, there is a further conversion of these organs into instruments connected with the mouth; so as to bring them nearer to the nature and use of maxillæ or under jaws, and of a labium or under-lip.

It appears from the experiments and observations of Rathke[3] that the long-tailed Decapod Crustaceans do not change the form, or increase the number of locomotive organs, that distinguish them when they issue from the egg.[4]

---

[1] *Cancer Pagurus.*     [2] *Brachyuri.*

[3] Récherches sur le dévélopement des Ecrevisses. Abstract of *Ann. des Sc. Nat.* xix. 442.

[4] *Ibid.* 463.

Once residing a few weeks on the northern coast
of Norfolk, where the sea, at low water, retires
to a considerable distance from the high water
mark, I had an opportunity of witnessing the
proceedings of a species of crab very common
there,[1] and varying greatly in size, some, if my
memory does not deceive me, scarcely exceeding
the size of a pea, others being three or four
inches in diameter, and all exactly corresponding in every particular; so that it seems probable that the short-tailed tribe also undergo no
change, except of size, though, as we have seen
above, the terrestrial Isopods acquire additional
legs in their progress to maturity. The legs,
however, of these Crustaceans cannot be regarded as analogues of the legs of *Hexapods*, but
rather of the *acquired* legs of the *Myriapods*.

In order to form a clear notion of the object of
Providence in thus, as it were, taking certain
organs from locomotion, and forming a new set
for that purpose, and multiplying those connected with the seizing and mastication of the
food of the animals in which this metamorphosis
takes place, it would be necessary to watch their
proceedings in their native element, the water,

---

[1] *Cancer Mœnas.* L. Mr. Westwood, in a letter received
since this went to press, expresses his conviction that Crustaceans
do *not* undergo any metamorphosis. Besides a variety of other
arguments which he will himself bring forward in due time, he
lately met with young specimens of this crab at Conway, in N.
Wales, only $\frac{1}{16}$ of an inch in length, which did not differ from
adult ones.

to ascertain the nature of their food, their mode of taking it, and other circumstances connected with its conversion into a pulp proper for digestion; but as few can have an opportunity of doing this, we can only conjecture that this multiplicity of organs is rendered necessary by the circumstances in which they are placed, and the element they inhabit; for, as we have seen, no such conversion occurs in the *terrestrial* Crustaceans; probably the denser medium requires a more complex structure and more powerful action in the instruments connected with the nutriment of the animal.

Having considered these instances of the *legs* of Hexapods being, as it were, metamorphosed into organs more especially connected with nutrition, I shall next mention, more briefly, some cases in which the oral organs themselves are modified to discharge *other* functions than what is usually their primary one.

To begin with the *Arachnidans* or spiders. In these the two-jointed *mandibles* or chelicerae, as Latreille calls them, are not organs of mastication solely; for though, from the vast strength and power of the first joint and its flat internal surface, we may conjecture that it assists in pressing the juices out of their prey, yet at the extremity of the second is a poison fang, being furnished, like the tooth of a viper or centipede, with a pore for emitting venom, which though not easily discovered in the smaller species, is visible under a lens in the larger; with these fangs, which communicate with a poison vesicle, the spider dispatches the insects struggling in his toils, which otherwise he could not so easily master, and having sucked out their juices casts away the carcase. The fang, by folding upon the apex of the basal joint of the organ we are considering, which is toothed on each side, and has a channel to receive it when unemployed, can be formed into a forceps, resembling that which arms the anterior thoracic leg of the shrimp, or that of the mantis, and which is probably, in some circumstances, used for prehension.

The subject of *poison-fangs* affords a striking example of the adaptation and modification of different parts and organs to the discharge of the same or similar functions, according to the circumstances in which an animal is placed; the viper, the centipede, and the spider have their sting in their *mouth*, or in its vicinity; the scorpion and the bee and wasp have it at the *other extremity* of the body; while the male of the *Ornithorhynchus*, or Duck-bill, and *Echidna*, or New Holland Porcupine, have it in their *hind legs*. Considering the evident affinity between these last animals and the *birds*, their poison-spur seems evidently analogous to the spur that distinguishes the males of many gallinaceous birds; and, reasoning from analogy, we may conclude that this organ is given to the males of the *Monotremes* as a weapon to be used in their mutual combats.

Whoever examines the underside of a spider will find the feelers and the eight legs arranged nearly in a circle, with their first hip-joints parallel; with some this joint in the feelers is dilated, but in others it is of the same shape with the analogous joint of the legs, only a little longer. It forms the *maxilla* or under-jaw, and between the first pair is the under-lip. The function of the maxillæ is to assist the, so called, mandibles, in pressing out the juices of the flies and other insects submitted to their action, and the analogous and parallel joints in the eight legs add some momentum to it.

The *Palpi*, or feelers—which in some cases emerge from the side of the maxilla, and appear a distinct organ, and in others are merely a continuation of it—in one sex undergo a singular conversion, and discharge a function connected with reproduction; and in the other, the female, are said sometimes to assist in supporting the egg pouch, which many of these creatures carry about with them, and guard with maternal solicitude.

It has been made a question by physiologists what the mandibles, and maxillæ with their palpi, of the Arachnidans really represent;

whether they are the analogues of organs bearing the same name in Hexapod Condylopes, or of others to be found in the Crustaceans or Myriapods. Latreille, in his latest work, regards the pieces immediately following the upper lip as analogues of the same parts in the Crustaceans, namely, a pair of palpigerous mandibles, two pairs of pediform maxillæ, and two pairs also of maxillary feet, analogous to the four anterior feet of insects.[1] Of the above organs, the mandibles and two pairs of maxillæ may be regarded as having their prototype in the Hexapods; for the second pair of maxillæ of the Crustaceans, in the Chilognathans, is the piece that represents the labium, or under-lip, of the first named animals.

Savigny, however, is of opinion that the auxiliary *maxillæ*, or, according to Latreille, maxillary *feet*, of the crab, except the first pair, become the *mandibles* and *maxillæ* of the spider; and that the *thoracic* legs of the same animal, with the same exception, become also its *ambulatory* legs :[2] thus accounting for the reduction of the number of the latter from *ten* to *eight*, perhaps he was induced to adopt this opinion, with respect to the oral organs, by considering the mandibles of the spider as analogous to the poison-

[1] Latr. *Cours D'Entomologie*, 167.
[2] *Anim. sans Vertébr*. ii. 57, Note *a*.

fang which arms the second pair of auxiliary feet of the *Scolopendra*.

I feel, however, rather inclined to adopt the opinion of the former learned entomologist, from the consideration of an *Arachnidan*, which seems evidently to lead towards the Hexapods. The animal I allude to is one of ancient fame, of which, once for all, I shall here give the history.

Ælian relates that a certain district of Æthiopia was deserted by its inhabitants in consequence of the appearance of incredible numbers of scorpions, and of those *Phalangians* which are denominated *Tetragnatha*, or having four jaws. An event mentioned also by Diodorus Siculus and Strabo.[1] Pliny likewise alludes to this event, but calls the last animal *Solpuga*,[2] a name which, in another place,[3] he says was used by Cicero to designate a venomous kind of *ant*.

The epithet *Tetragnatha*, applied by Ælian, &c. to the animal which, in conjunction with the scorpion, expelled the Æthiopians, as just stated, from the district they inhabited, seems clearly to point to the Solpuga of Fabricius, for any person, not skilled in natural science, would, when he saw the expanded forceps of their mandibles, pronounce that they had *four*

[1] Bochart. *Hierozdic.* ii. l. iv. c. 13.

[2] *Hist. Nat.* l. viii. c. 29. This name seems derived from the Greek, *Heliocentris*.

[3] L. xxix. c. 4.

jaws;[1] and the animals of this genus, in their general form and aspect, exhibit no small resemblance to an *ant*, so that it is not wonderful that Pliny should regard them as a kind of venomous ant. It seems, therefore, almost certain that the ancient and modern Solpuga are synonymous. Pliny, indeed, mentions a certain kind of spider—one of which he describes as weaving very ample webs—under the name *Tetragnathii;* but these appear to have no connexion with the *Phalangia tetragnatha* of Ælian, &c.

Olivier was the first modern naturalist who described the animals now before us, to which he gave the generic appellation of *Galeodes;* but if, as the above circumstances render very probable, they are really synonymous with the ancient *Solpuga*, that name, revived by Fabricius, should be retained.

Whether these animals are really as venomous and maleficent as they were said to be of old, and as their terrific aspect may be thought to announce, seems very doubtful. We learn from Olivier that the Arabs still regard their bite as mortal, and that the same opinion obtains in Persia and Egypt; and Pallas relates several facts, which, he says, he witnessed himself, which appear to prove that, unless timely remedies are applied, they instill a deadly venom into those

---

[1] L. Dufour. *Annal. Génér. des Sc. Nat.* iv. *t.* lxiv. *f.* 7, *a.*

they bite. Oil is stated to be the best application. On the other hand, Olivier, who found
these Arachnidans common in Persia, Mesopo-
tamia, and Arabia, affirms that every night they
ran over him, when in bed, with great velocity,
without ever stopping to annoy him; no one
was bitten by them, nor could he collect a single
well-attested fact to prove that their bite was
so dangerous: to judge by the strong pincers
with which the mouth is armed, he thought it
might be painful, but he doubts whether it is
accompanied by any infusion of venom. The
mandibles have clearly no fang with a poison-
pore, like those of the spiders.

To return from this digression. I principally
mentioned this tribe of animals, because, as was
long ago observed by Walckenaer,[1] and the observation was repeated by L. Dufour,[2] the head,
in them, is distinct from the trunk; and, as well
as *Phrynus* and *Thelyphonus*, it has only six
thoracic legs: so that, as the latter writer remarks, though its physiognomy and manners
arrange it naturally with the Arachnidans, these
characters exclude it from them.[3] Latreille,
indeed, seems to regard the head and trunk of
this animal as not distinct, but as forming together what he names a *cephalothorax*, or head-
thorax; yet he admits that the three last pairs

---

[1] *Tableau des Araneid.* 1.      [2] *Ubi supr.* 18.
[3] *Ibid.* 20.

of legs are attached to as many segments of the trunk,[1] which certainly infers the separation above alluded to.

Savigny says, with respect to the feelers of *Solpuga*, that they, and the two anterior legs, so closely resemble each other, that they may either be called feelers or legs; but in the species described by L. Dufour,[2] and another in my cabinet,[3] this is not altogether the case, for the feelers, though pediform, are not terminated by a claw, but by a membranous vesicle, from which issues, when the animal is irritated, an apparatus probably used as a sucker, and which gives them a prehensory function; while the organs that represent the anterior pair of legs of the other Arachnidans, at the base of their maxillary or sciatic joint, are soldered, as it were, to the corresponding joint of the feelers, with which they agree in the number and kind of their articulations, except that they do not protrude a sucker; neither are they armed with a claw like the other legs, but are probably simply *tentacular*, or exploratory. There seems no slight analogy between these united maxillæ and what Savigny denominates the first and second pair of maxillæ of the millepedes, also united, which appear to me to represent the lower-lip and maxillæ of the hexapods, and in

---

[1] *Cours D'Entomolog.* 548.     [2] *Galeodes intrepidus.*

[3] *Solpuga fatalis.*

this case the two pair of feelers that issue from the coxo-maxillæ, as they are sometimes called, or sciatic joints in the *Solpuga*, may be regarded as representing the *labial* and *maxillary* feelers of the hexapods; the second pair are also analogous, both in their place and their function, to the first pair, or tentacular legs of *Thelyphonus* and *Phrynus*. In the *Solpuga*, the labium, or under-lip, of the spiders, is represented by a bilobed organ, which Savigny calls a *sternal* tongue.

From the consideration of this animal we seem to have obtained the elements, or type, in reference to which the oral, prehensory, and locomotive organs of the Arachnidans were formed; that their mandibles, maxillæ, and feelers; their second maxillæ, and the, so called, anterior legs emerging from them, are analogous to the mandibles, labium and labial feelers, and maxillæ and maxillary feelers of the hexapods; and the remaining three pairs of legs, of their six legs; the sternal tongue, so called by Savigny because it is a process of the sternum, will thus be an organ *sui generis*, unless it may be regarded as, in some sort, the analogue of the prosternum of insects. If this view is correct, we have here various conversions, as of *maxillæ* and *palpi* into *legs;* a *labium* into *maxillæ;* and a *prosternum* into a *labium.* In the *Pedipalps*— with the exception of the scorpions,—*e. g.* in

*Thelyphonus* and *Phrynus*, especially the latter, the *first* pair of legs of Octopods seem to wear the form, and in some measure to discharge the functions of *antennæ*.

In the *shepherd-spiders*[1] *all* the legs, in some degree, imitate antennæ, especially in their *tarsi*, which sometimes consist of more than *fifty* joints, rendering them very flexible, so as to assume any curve, and fits them, as their long legs do the *crane-fly*,[2] to course rapidly over and among the herbage and the leaves of shrubs, &c. When reposing upon a wall, or the trunk of a tree, this animal arranges its legs so as to form a circle as it were of rays around the body, the thigh forming a very obtuse angle with the rest of the leg, and so, though the body is so small, they occupy a considerable space; but, if a finger, or any insect, &c. touches them, it elevates these angles into very acute ones, so as to form a circle of arcades round the central nucleus or body, under which any small creature can pass, but if this does not succeed, it makes its escape with a velocity wonderful for an animal furnished with legs more than ten times the length of its body.

In the *scorpion* and the *book-crab*,[3] as well as the shepherd-spider, the mandibles, which are short, have a moveable joint, and are converted into a forceps, like the anterior legs of the crab or

---

[1] *Phalangium.*    [2] *Tipula.*    [3] *Chelifer, Obisium,* &c.

the lobster; their feelers also, which are very long, terminate in the same way, and form an organ by which they can catch their prey; the former being armed besides with a long jointed tail, furnished at the end with a sting, which they can turn over their back, and thus, either annoy their assailants, or dispatch any captive whose resistance they cannot otherwise easily overcome.

To what a variety of uses are analogous organs applied in the diversified instances here adduced; and in all these variations from a common type, how apparent are the footsteps of an intelligent First Cause, taking into consideration the intended station and functions of every animal, and how the structure may be best adapted to them, not only in general, but in every particular organ.

As far as we can lift up the mystic veil that covers the face of nature, by means of observation and experiment, we find that every iota and tittle of an animal's structure, is with a view to some end important to it; and the Almighty Fabricator of the Universe and its inhabitants, when he formed and moulded, *ex præjacente materia,* the creatures of his hand, decreed that the sphere of locomotive and sentient beings should be drawn together by mutual attraction, and concatenated by possessing parts in common, though not always devoted to a common use; thus leading us gradually from one form to another, till we arrive at the highest and most distinguished of the visible creation ; and instructing us by his works, as well as by his word, to cultivate peace and union, and to seek the good of the community to which we belong ; and, as far as our influence goes, of the whole of His creation.

## Chapter XVII.

### Motive, locomotive, and prehensory Organs of Animals considered.

THE remarkable circumstances noticed in the last chapter with regard to the legs of Crustaceans and Myriapods, and their employment in aid of manducation, sheds no small light upon the subject of locomotive organs in general, and their primary function ; it will therefore not be out of place, if, in the present chapter, I consider those organs, as far as they are *external*, according to their several types, as exhibited in the entire sphere of animals ; upon which, indeed, the due accomplishment of their various functions, and the exercise of their several instincts—which in most of the succeeding classes assume a new and more developed character—mainly depend. This is a wide field, but one full of interest, and which, studied as it deserves, conspicuously illustrates the higher attributes of the Deity.

We are placed in a world full of *motion*; of all motions, none fall more immediately under our notice than those of the various members of the animal kingdom ; and the external organs by which they are effected, attract every eye both by their infinite diversity, and the ,adaptation of their individual structure to the occasions and wants of the animal in whom they are found, so that they may, in the best and safest manner, effect such changes of place as are necessary for their purposes.

Nutrition may be stated as the primary object of the motions and locomotions of the members of the animal kingdom in general. No sooner is the fœtus or embryo so separated from its parent stock, as not to imbibe its food from it, than it begins to employ instinctively its prehensory and motive organs in collecting it. And, whether we descend to the foot of the scale of animals, or mount to its summit, we shall find that their —*Daily Bread*—is the principal object that in every Class sets the members in motion.

The *motive* organs may be divided into *two* classes, those that are employed by an animal in *locomotion*, and those that are used for *prehension* ; but as many of the locomotive organs are

also prehensive, and prehension is often in aid
of locomotion—as in climbing and burrowing—
it will not be easy to consider the motive organs
separately with regard to these functions, I shall
therefore consider them generally, according to
certain types or kinds, under which they may
be arranged, and which present themselves very
obviously, when, with this view, we survey from
base to summit, or rather from pole to pole,
the entire sphere which constitutes the animal
kingdom.

Generally speaking, in this survey, as well as
in the peculiar motions of the various groups of
animals, we have no trouble in ascertaining what
are the external organs by which the Creator has
enabled and instructed each animal to accomplish
them; but there is one anomalous tribe, or,
perhaps, it might be denominated, *Sub-kingdom*,
in one Class of which, at least, this is not so
obvious. I allude to Ehrenberg's Tribe of *Plant-
animals*,[1] particularly his first or polygastric
Class,[2] in which the organs of their various locomotions, enumerated in a former part of this
work,[3] remain unknown, and some, as those that
have an oscillatory movement, one might almost
suspect were moved by an *external* cause. The
little Monad, parasitic on the eye-worm of the
perch,[4] which alternately spins round like a top,

---

[1] *Phytozoa.*      [2] See VOL. I. p. 156.
[3] *Ibid.* 153.      [4] *Diplostomum volvans.*

and then darts forward like an arrow,[1] seems as if, like a watch, it required to be wound up before it could go.

Before I confine my observations to those motive organs which are local and planted in certain parts of the body of an animal, as legs, wings, fins, &c., I shall first mention those motions in which the whole body is concerned. Of this description is the alternate expansion and contraction of some, as the Salpes and Pyrosomes and other Tunicaries;[2] the annular motion propagated from one extremity of the body to the other, as in the earth-worms,[3] geometric caterpillars, and many other larvæ; the undulating movements of the flexile bodies of many aquatic animals, as fishes, particularly the serpentiform ones; and the gliding motion of serpents themselves over the surface of the earth as well as their undulations. Many of the animals here alluded to are provided with subsidiary organs— as the earth-worm with lateral bristles;[4] the geometric larvæ, with legs at each extremity of their body; the leach with suckers; which, however, would be of little use without the expansion and contraction of its body;[5] and the fishes with fins: but if we consider the form and

---

[1] VOL. I. Appendix, p. 354.  [2] See VOL. I. p. 223, 227.
[3] *Ibid.* p 340.  [4] *Ibid.*
[5] *Ibid.* p. 336.

circumstances of all these animals, we shall see, in each case, the design and contrivance of Supreme Wisdom. Without the power of contraction and expansion, by which the Salpes, Pyrosomes, &c., alternately attract and repel the waters which they inhabit, they might indeed, from their absorbent structure, be saturated, but nutrition could not take place. The earthworm again, a subterranean animal, but which occasionally emerges, by the annular motion of its body can much more easily wind its sinuous way without obstruction when it seeks again its dark abode under the earth. The denser medium compared with air, through which the aquatic animals pass, renders great flexibility a very important quality, to enable them to overcome the resistance it opposes to their progress.

Having premised these observations on motions produced by the action of the whole body, or successively propagated from one extremity to the other, I shall now proceed to consider those external organs, which are its obvious instruments in the great majority of animals, beginning with those that are found in the *lowest* groups.

1. *Rotatory Organs.* In some species of Infusories, even in Ehrenberg's first Family of his polygastric Class, the oral aperture is *fringed* with a circlet of bristles, but whether the animal by their means creates a vortex in the water, or whether they are analogous to the tentacles of

the polypes, and are employed in collecting its food, seems not to have been clearly ascertained. Lower down in this Class, and approaching the Rotatories, we find a singular animal,[1] with bristles, by their position, simulating legs, which, as was before observed,[2] revolve with wonderful rapidity. But it is in the Class of *Rotatories* that these revolving organs are most conspicuous. They are described as shaped like a tunnel, the tube of which terminates in a deep-seated pharynx armed with jaws, and the external dilated orifice fringed with fine hairs or bristles, to which the animal communicates a very rapid rotation, whence they are called *wheel-animals*. Some, as the vorticels,[3] the wheel-animals by way of eminence, appear to have *two* wheels, others *three*, or even *four*: Lamarck is of opinion, from the observations of Du Trochet, that what are taken for two or more wheels, are only one, bent so as to form partial ones;[4] but in some they are certainly distinct organs.[5] The object of the rapid gyration of this wheel or wheels is to

---

[1] *Discocephalus Rotator*, PLATE I. A. FIG. 6.

[2] VOL. I. *Appendix*, p. 350.

[3] *Vorticella*. Müll. They constitute chiefly the *Rotifera* of Lamarck, and are divided by Ehrenberg into numerous genera. His genus *Vorticélla*, the type of which is *V. convallaria*, Müll. is placed in his Polygastric Class, in a section of his fourth Family (*Anopisthia*), which section he names *Vorticellina*.

[4] See Baker *On the Microscope*, i. 91. *t.* viii. *f.* 5.

[5] *Ibid. f.* 6.

create a vortex in the water, whose centre is the mouth of the animal, a little charybdis bearing with it all the animalcules or molecules that come within its sphere of action, and by this remarkable mechanism it is enabled by its Creator, as long as it is encircled by a fluid medium, to get a due supply of food. These wheels are merely foraging organs, for on a surface the locomotions of these singular animals resemble those of the leech described in another place.[1]

In surveying the organs by which animals procure their food, we are struck by the wonderful diversity and multiplicity of means by which the same end is attained, and yet, through all this diversity, a series of approximations may be traced, proving that the same hand directed by the Wisdom, Power, and Love of one and the same Infinite Being fabricated the whole host of creatures endowed with powers of voluntary motion. What care does it manifest, and attention to the welfare of these invisibles, and what contrivance, that they should be fitted with an organ, by means of which, when they are awakened from a state of suspended animation, and from a long fast perhaps of months, or even years, by water coming in sufficient contact with them, they can start up into life, and by the gyrations of their wheels immediately begin to breathe, and to procure a sufficient supply of

[1] See Vol. I. p. 336.

food for their sustenance, while they continue animated.

2. *Tentacles.* Nearly related to these bristle-crowned rotatory appendages of the mouth of some animalcules are what are named Tentacles. so called probably from their being usually exploring organs. In its most restricted sense, this term is understood to signify organs, appendages of the mouth, which have no *articulations,*[1] but, in a larger sense, the term has been applied also to all jointed organs in its vicinity, and used for a similar purpose, which indeed are the precursors of feelers and antennæ. The structure of the first-mentioned, or proper tentacles, and the means by which they perform their motions, and fulfil their functions, have been before explained.[2] It is to these organs, as well as for their food, that the polypes are indebted for what constitutes their principal ornament, that resemblance which, though born to blush unseen, even in the depths of the ocean, their Creator has enabled them to assume, of a plant or shrub in full blossom adorned with crimson or orange-coloured flowers.

In the *fixed* polypes, the tentacles are the only motive organs, but in those that can *shift their quarters,* as the *Hydra,*[3] they move by fixing each extremity like the leech, probably

---

[1] See Savigny *Syst. des Annelides,* iii. 4.
[2] See Vol. I. p. 164.          [3] *Ibid.* 173.

by means of something analogous to suckers.
As the former, like their analogues in the vegetable kingdom, are fixed by their base, and
consequently cannot move from place to place
in search of food, Divine Goodness has compensated this to them, and they obtain all the
advantages of locomotion by the progressive
multiplication of their *oscula* or mouths, each
surrounded by a coronet of tentacles, so that
they have, on all sides, and at all heights, numberless sets of organs constantly employed in
collecting food from the fluid they inhabit; some,
it is stated, by creating a vortex, like the wheel
animals, and the majority, probably, by means
of minute suckers, or some viscid tenacious
secretion. What each individual collects does
not merely serve for its own nutriment, but also
contributes something to that of the whole community,[1] so that though some may contribute
more to the common stock and others less, yet
the deficiency of one is made up by the redun-
dancy of another.

The tentacles of the fresh-water polypes
forming the locomotive genus *Hydra*, are not, as
those of the fixed marine ones, shaped like the
petals of a blossom, but are long hair-like flexile
arms, somewhat resembling the branches of a
chandelier,[2] which explore the waters around

---

[1] See Vol. I. 171.

[2] Lasser. L. *Théologie des Ins.* i. t. ii. *fr.* 28—32.

them, and lay strong hold of any small animals or substances they come in contact with,[1] so that they seem to throw out lines, fitted with hooks, to catch their prey.

Amongst the *Radiaries*, in the Order of *Gelatines*,[2] tentacles exist in some genera and not in others, and, where they do exist, their functions and situation are not clearly ascertained. In the *Pelasgic Medusa* there are four broad flexible arms, and round the margin eight narrow tentacles, as they are called, both of which the animal is stated to employ in seizing its prey, so that both may be entitled in this view to the denomination of *tentacles*, yet one may be respiratory organs and the others merely prehensory.[3] But the Medusidans vary greatly with regard to these organs, some having neither arms nor tentacles;[4] others having tentacles but no arms;[5] others again arms but no tentacles;[6] and lastly, others both these organs.[7]

In the two first sections of the Order of Echimoderms, consisting of the *Stelleridans* and *Echinidans*, the mouth has no coronet of tentacles, but, instead, is armed with five pieces, which, in the latter particularly, assume the form and function of *mandibles*;[8] but the Fistulidans

---

[1] See VOL. I. 165—170.  [2] p. 195.
[3] Carus. *Comp. Anat.* i. 47.  [4] *Eudora.* Lam.
[5] *Equorea.* Lam.  [6] *Cassiopea.* Lam.
[7] *Aurelia.* Lam.  [8] PLATE III. FIG. 9—11.

present again a floriform coronet of tentacles, not simple but expanded, and branching at their extremity, with which they seize their prey. In the *Holothuria*, besides these, the mouth is armed with five teeth or mandibles.

Tentacles, but not conspicuously, surround the mouth of only some of the *Tunicaries*, it will therefore be sufficient merely to mention them, and proceed to certain oceanic animals amongst the *Annelidans* whom their Creator has adorned, if I may so speak, with rays of glory, which, when expanded, surround their head, or rather mouth, with a most magnificent coronet. The animals I allude to constitute the genus *Amphitrite* of Lamarck, and the *Sabella* of Savigny; this coronet, in some species, is formed by numerous tentacles, called, by the authors just named, *Branchiæ*, or gills; but as they are stated to be employed in collecting their food, as well as in respiration,[1] they seem in this respect perfectly analogous to the tentacles of the polypes, and wheels of the rotatories, which are also respiratory organs. The great difference seems to consist in their being divided into two fan-like organs in the Amphitrites, in which the digitations or tentacles proceed from a common base, and which together form the coronet. In some the digitations, like the sticks of a fan, are

---

[1] Lamarck, *Anim. sans Vertebr.* v. 355.

connected by an intervening membrane, thus resembling two expanded fans;[1] in others, this pair of organs forms two bunches, set, as it were, with numerous spirally convoluted plumes;[2] in a third each bunch of plumy tentacles is convoluted, but not spirally;[3] but the most magnificent species of the genus, if indeed it belongs to it, is that figured in the fifth volume of the *Transactions of the Linnean Society*,[4] under the name of *Tubularia magnifica*. I say, *if indeed it belongs to it*, because, if the figure quoted is correct, which I am not aware there is any reason to doubt, the gills or tentacles, call them which we will, are not, as in the other species, divided into two fasciculi or bundles, the rays of which sit upon a common base; but form one glorious and radiant coronet, whose rays are beautifully annulated with red and white; there appears indeed to be a double circle or series of these rays, the interior ones shorter than the exterior; but there is not the least appearance of their division into *two* bunches, each forming a semicircle. The rays differ little from those of many of the polypes, except in being more numerous and longer, for the diameter of the circle, when the rays are all expanded, is nearly six inches, and it is not stated that the figure is magnified.

---

[1] *Amphitrite Infundibulum. Linn. Trans.* ix. *t.* viii.

[2] *A. volutacornis. Ibid.* vii. *t.* vii. *f.* 10.

[3] *A. vesiculosa. Ibid.* xi. *t.* v. *f.* 1.  [4] *Ibid. t.* ix. *f.* 1—5.

Whenever the animal is alarmed it withdraws this gorgeous apparatus of respirato-prehensory organs within its tube, and the tube itself into its burrow in the living rock, as a safe refuge from its enemies. Whoever compares the above figure of this expanded animal-blossom with the nectaries of some species of passion-flower, will be struck by the resemblance they exhibit to each other,[1] and by the analogy that evidently exists between them. As prehensory organs, the principal object of their unusual length and numbers may probably be their capturing, as in a net, a quantity of rock animals, or animalcules, sufficient for their support, and perhaps their very beauty may be a means of attraction and bring them within their vortex.

With these splendid animals we bid farewell to those whose oral organs seem analogous to the blossoms of vegetables, and also to those in whom the organs of prehension and respiration are united; or in which the same organs collect food and also act the part of gills.

Though tentacles are not henceforth employed in *respiration*, yet they still exist in several other classes of animals as exploratory, prehensory, and locomotive organs. But in none are they more remarkable, both for their structure and uses, than in the Cephalopods or cuttle-fish. In

[1] See LINN. TRANS. ii. *t.* iii. *f.* a. b.

these animals they are used, as we have seen, as arms for prehension, as legs for locomotion, as sails for skimming the surface of the ocean, as oars for passing through its waves, as a rudder for steering, and as an anchor to fix themselves.

These organs, like the tentacles of the polypes, surround the mouth; in some genera, as the poulpe,[1] and sepiole,[2] besides eight shorter arms,[3] there is a pair of very long ones, which are usually denominated tentacles, by way of eminence, which the animal probably uses, and for which purpose a claw arms their extremity,[4] to lay hold of prey at a distance. The means by which the tentacles perform the locomotions of these animals, and enable them to seize their prey, I shall advert to under another head.

But though, in the great body of the Cephalopods, the tentacular organs do not exceed *ten*, we find, from Mr. Owen's admirable memoir on the *Pearly Nautilus*,[5] that, in that animal, they are extremely numerous, and strikingly different in their structure. The mouth and its appendages are retractile within the head, which forms a sheath for them, the orifice of which is anterior. The proper tentacles are of two kinds: 1. Brachial ones, finely annulated, emerging from thirty-eight three-sided arms, disposed ir-

---

[1] *Octopus.*            [2] *Sepiola.*
[3] PLATE VII. FIG. 3. *a.*      [4] *Ibid. b.*
[5] *Nautilus Pompilius.*

regularly, nineteen on each side, all directed forwards, and converging towards the orifice of the oral sheath. 2. Labial ones, similar to the others in their structure, and emerging from four broad flattened processes, arising from the inner surface of the sheath, and more immediately embracing the mouth and lip: from each of these processes emerge twelve tentacles, rather smaller than the brachial ones. Besides these two descriptions of tentacles, there is a pair, one on each side, emerging from two orifices in the inner part of the hood or foot, arranging with the arms, and perhaps to be reckoned with the brachial tentacles, thus making up the whole number of tentacles of a similar structure eighty-eight. It is to be observed that neither the parts that sheath them, nor the tentacles themselves, are furnished with any acetabula or suckers.[1]

Besides the tentacles, this animal has four analogous organs of a different structure, one before and one behind each eye, which Mr. Owen likens to antennæ, and which are lamellated, or composed of a number of flattened circular disks, appended to a lateral stem;[2] a circumstance indicating a variation in their functions.

From their being retractile, it should seem

---

[1] Owen's *Memoir*, &c. 13, *t. i. n.*       [2] *Ibid.* 14.

that in this animal the tentacles are not in constant use, as they are in the naked Cephalopods, and that they require protection; from their finely annulated structure they appear to be flexible and easily applicable to any surface, but whether they are tentacular or prehensory organs, or both, is unknown. In the account of the *Loligopsis,* a species of cuttle-fish, by the able pen of that eminent zoologist Dr. Grant, the part apparently analogous to the labial tentaculiferous processes of the Nautilus, is called the *outer-lip,* and is stated to send out a muscular band to the base of each *arm,*[1] which seems to indicate that the arms of the naked Cephalopods are analogous to the labial tentacles of the animal we are considering. The labial processes, with their tentacles, present some resemblance to a many-fingered hand,[2] and from their situation immediately next the mouth may be conjectured to be most concerned either in the capture or transmission of its food: but whether either set of tentacles is used in its locomotions, as they are in the naked Cephalopods and the Argonaut, seems very problematical.

As far as its locomotion on a surface is concerned, in its hood, it appears to be furnished with an expansile foot, approaching that of the

[1] *Trans. of Zool. Soc.* I. i. 23.
[2] Owen, *ubi supr. t.* iv. *f, i i, g g.*

*Gastropods*,[1] so that its tentacles seem not necessary to transport it from place to place on the bed of the ocean; by what means it elevates itself, as it is known to do, to the surface, and floats upon the waves, has not been ascertained.

In comparing the organs that surround the mouth of the Nautilus with those of other Cephalopods, we see that a vast change has taken place. They are no longer the principal organs of locomotion, that function being transferred to an expansile foot; their number is increased in nearly a tenfold ratio: being deprived of suckers, they seem destitute of any powerful means of prehension and retention, and so are scarcely able to overcome the resistance of the larger Crustaceans. As their principal organ of locomotion is one that seems to preclude all idea of rapid motion in pursuit of their prey, it is most probable, as their mandibles are fitted for crushing crust or shell, that certain Molluscans, animals which must be equally slow in their motions, and can scarcely resist them, are their destined food.

We may further observe, that, regard being had to the organs which surround the mouth, a very wide interval separates the great body of the Cephalopods, known in a recent state, from the animal now before us; even the *Spirula*, which Mr. Owen conjectures may belong to the

[1] Owen's *Memoir*, &c. 12, *t.* i. *n.*

same Order, in this respect is formed upon a very different type, precisely that of those Cephalopods.[1]

This animal, in the above respect, being so completely insulated, it seems, as if in its means of entrapping its prey it was formed upon a plan not connected with that of any other Molluscan, but quite *sui generis:* probably, were we acquainted with the animals belonging to what are deemed fossil Cephalopods, we should find the hiatus vastly narrowed.

In this instance we see clearly that adaptation of means to an end which distinguishes all the works of the Creator; the striking variation which this creature exhibits from the oral apparatus of its Class, is evidently connected with the kind and circumstances of the animals which it is commissioned to keep within their proper limits; its mandibles, or beak, indeed, resemble those of the other Cephalopods, indicating that its prey are covered with solid integuments, requiring great force to crush them; but the other oral organs, and its snail-like foot, as we see, indicate that they are not of a kind that can easily escape from their assailants.

Two objects seem to have been principally in the mind of the Almighty planner of the universe of beings: one seems to have been the concatenation of all subsistences, seriatim and

[1] PLATE IV. FIG. 2.

collaterally, into one great system; and the
other, so to order and vary the structure of each
individual that it may be duly fitted to answer a
certain end, and produce a certain effect upon
such and such points of that system, and this in
such a way that these effects, though *diverse*,
might not be *averse*, but proceed, if I may so
speak, in the same direction. Thus, in the
subject before us, the general commission given
to the Cephalopods, is to assist in reducing
the *armed* population of the ocean within certain
limits, and to all are given instruments and
organs, varying indeed in their structure, but
proper to enable them to effect this purpose;
all, however, concurring to bring about a common
and connected object, and one taking one department and another another.

The tentacles of the *Univalve Molluscans*, for
the headless animal of the *Bivalves* has no such
organ, are neither used for locomotion nor prehension, and therefore seem to have no claim
to a place in the present chapter. But as they
are clearly the analogues of the tentacles of the
animals we have been considering, and though
not prehensory, are certainly exploring and sensiferous organs, which are probably connected
with prehension, I shall make a few observations
upon them. They vary in their number, some
having none,[1] others only two;[2] others again

[1] *Chiton.*    [2] *Cypræa. Voluta.* PLATE VI. FIG. 1. *b.*

four;[1] and lastly, others six.[2] They are without articulations, though they sometimes exhibit an annulated appearance:[3] they are also often retractile, and in the snail and slug they form a hollow tube, which can be inverted like the finger of a glove; in others they appear to be composed of longitudinal fibres, intersected by annular ones, which render them capable of great extension. In form they are either filiform, setaceous, or conical; but in the remarkable genus *Laplysia*, or the Sea-hare, the upper pair are shaped like the ears of the animal from which they take their name. Their sense of touch is much more delicate than that of the rest of the body. They are intimately connected with what are usually deemed the organs of sight of the Univalve Molluscans, which in some genera they seem to inclose. Some of these eyes are placed, in the form of a black pupil, at the summit of the tentacle, which surrounds them as the iris does the pupil of the perfect eye; in others they are imbedded in the middle of that organ, and in others at its base; in some, as in the Sea-ear,[4] they are seated in a separate footstalk. In many of the carnivorous species the pupil is

---

[1] *Helix. Limax.*

[2] *Clio.* The tentacles in this genus are retractile, and when retracted form two tubercles, which make the head appear bilobed.

[3] *Voluta Æthiopica,* PLATE VI.　　　[4] *Haliotis.*

surrounded by an iris,[1] which seems to indicate
that the tentacles perform, in some sort, the
functions of that part of the eye. The upper
pair of tentacles in the Molluscans seem analogues of the *antennæ* of Condylopes, and the
lower pair of their *feelers;* and the functions for
which the Creator has formed and fitted both
are probably not very dissimilar. The extreme
irritability of the tentacles of snails and slugs is
evident to every one who observes their motion :
at the approach of a finger they are immediately
retracted ; they therefore give notice to the animal of the approach of danger, so as to provide
against it, and when necessary to withdraw
itself into its shell : the eyes, from their situation
in many of them, supposing them to have a greater
range and power of vision than they appear to
have, cannot direct them in the choice of their
food, in these their lower tentacles may have this
office. Snails and slugs, we also know, issue
forth from their places of concealment when the
earth is rendered moist enough, by showers,
for them to travel easily over its surface ; so
that they must be endued with some degree of
*aëroscepsy,* of which probably these delicate
organs are the instruments.

Whether the barbs appended to the mouths of
many fishes, as the barbel, the Siluridans,[2] and

---

[1] PLATE VI. FIG. 1, *a.*        [2] PLATE XII. FIG. 1.

the Fishing-frog,[1] may be regarded as a kind of tentacle cannot be certainly affirmed, but from their proximity to the mouth, it seems most probable that they exercise some function connected with the procuring of its food. Cuvier regards them as a kind of tactors, and they also present some analogy to antennæ and palpi.

In many of the Annelidans, tentacles of the present description are found not only in the vicinity of the mouth, but also upon the pedigerous segments of the body, and appear to be equally used in exploring objects.[2]

I shall next consider some tentacular organs, which differ from those we have been considering in being more or less jointed. These, on that account, have been considered as a different class of organs, and by many have been denominated *cirri* or tendrils, or more properly, by Savigny, tentacular cirri. I have before described organs of this kind in my account of the *Cirripedes*,[3] by which it appears that they are employed for the same purposes as the tentacles of the polypes. Under this head also the antennæ of Crustaceans and insects may be noticed, which seem, as I have lately observed, analogous to the tentacles of the Molluscans, and the barbs of fishes; in some instances, indeed, they are

---

[1] *Lophius.* PLATE XIII. FIG. 2.
[2] *Fn. Groenland,* 294.     [3] See above, p. 2.

used instead of the fore legs.[1] The reason why
their structure differs from the soft, inarticulate
tentacles above described, at least in most cases,
appears to be the different nature of the integuments of the animal, which being incased in a
kind of coat of mail, it seems requisite that both
its locomotive and oral organs should be similarly
defended, and in this case, unless they had been
jointed, they would have lost their flexibility,
and so could not have exercised the functions
assigned to them by their Creator. It may,
perhaps, be objected that the shell of the snail is
nearly as hard as the crust of the lobster; but
when we consider that the former, when moving,
can thrust forth the greatest part of its soft body,
as it were from a house, while the crust of the
other is really its skin, this objection seems to
vanish.

*Suckers.*—The organs I am next to consider,
*acetabula*, or suckers, are, in many cases, so intimately connected with tentacles, as to form the
most essential feature of them, without which
they can be of no use. In fact, in the Cephalopods, they bear the same relation to the organ
just named that the hand or foot do to the arm
or leg, or the fingers and toes to the hand, in
higher animals : they are the part by which the

[1] *Introd. to Ent.* ii. 308.

animal takes hold of what it wants to seize ; and by the alternate fixing and unfixing of which, upon a solid substance, it moves from place to place. A sucker [1] may be defined—An organ by which an animal is enabled to create a vacuum between it, (the organ,) and any surface on which it rests, so as to produce a pressure of the atmosphere upon its upper part, and thus causing it to adhere firmly.

Cuvier, speaking of the suckers of the Cephalopods, thus describes their action. When the animal approaches one or more of its suckers to a surface, in order to apply it more intimately, it presents it flattened ; when it is fixed to it by the perfect union of the surfaces, it contracts its sphincter, which produces a cavity, in the centre of which a vacuum is formed. By this mechanism, the sucker attaches itself to the surface with a force proportioned to its diameter, and to the weight of the column of water or of air of which it is the base. This force, multiplied by the number of suckers, gives that with which the whole or part of the legs attaches itself to the body, so that it is more easy to tear the legs, than to separate them from the object which the animal wishes to retain. [2]

In some cases, the action of the suckers, as

[1] Suckers are denominated scientifically *Acetabula*, and *Cotylæ*, or Cotyloid processes.

[2] *Anat. Comp.* i. 410. Roget, *B. T.* i. 260.

suckers, seems not sufficient for the animal's purposes, and claws are superadded. This structure is to be found in the suckers of the animal that fixes itself to the gills of the bream, the *Diplozoon*, before described,[1] and to those of some Cephalopods a stout claw is added.

When we consider the nature and predatory habits of those Cephalopods whose tentacles are furnished with suckers, often pedunculated, on that side which is prone when the animal moves, we shall at once see the reason that this change from the more common Molluscan structure of an expansile foot, took place, for had their principal locomotive and prehensory organ been of this description, or different from what it is, their motions must necessarily have been so slow, and their powers of prehension so weak, that they could never have overtaken and captured, and maintained their hold of the well defended and formidably armed Crustaceans, which are their destined prey. Uncouth, therefore, and misshapen, and monstrous, as these animals, at the first glance, appear, we see that in these organs, and doubtless in all others, they are exactly fitted to answer the end, and fulfil the purposes of Divine Providence in their creation.

The suckers of the *Diplozoon* exhibit a com-

---

[1] Vol. I. Appendix, p. 358.

plex structure in aid of its powers of suction, not easily developed and understood. Dr. Nordmann supposes, that though the animal could attach itself strongly by these organs, additional means were necessary to render its attachment sufficiently firm; and that, therefore, while it is fixing itself by the suckers, it requires the aid of the apparatus of hooks, or claws and arches, to keep itself from being misplaced.[1]

The Class of *Annelidans* exhibits a great variety of locomotive organs, amongst the rest, in the last Order, we find *suckers*, these being the principal organs for motion of the *Hirudineans* or leeches, the animals of which Order, however, M. Savigny is disposed to think are essentially distinct from the rest of the Annelidans, on account of their want of *setæ* or lateral bristles. The *oral* sucker of that division of the animals I am considering, to which the common leech[2] belongs, is distinguished from the *anal* one by being formed of many segments, whereas the latter consists of only one. Their motions, by means of these suckers, and the annular structure of their bodies I have before sufficiently described.[3] Their suckers also enable them to lay hold of any aquatic animals that come in

[1] See *Nordmann*, i. 61. *t. v. f.* 3, 4, 5.

[2] *Sanguisuga medicinalis.* Sav.

[3] Vol. I. p. 336.

their way, especially the *oral* one, which once fixed they soon make an entry and begin to imbibe its blood.

We see, in this, the reason why their Maker, instead of bristles for locomotion, has given them organs by which they can not only move from one place to another, but also fix themselves firmly to their prey.

I shall next advert to a kind of sucker which really becomes both the hand and foot of the animals that bear them. I allude to those of the *Echinoderms*, described on a former occasion,[1] in which the ampullaceous part within the shell presents the first outline of a shoulder or thigh, the exerted extensile part that of an arm or leg, and the dilated part with which the animal seizes its prey or walks, the hand or foot; the two first constituting the tentacle, and the last the sucker.

I have, on a former occasion, given some account, under the name of the *Perch-pest*,[2] of a singular animal, belonging to the *Lerneans*, whose history has been given by Dr. Nordmann, and which is distinguished by a sucker common to *two* legs. Several other Lerneans have similar suckers.[3]

---

[1] See Vol. I. p, 202, 208. Plate III. Fig. 5.

[2] See above, p. 22, 31.

[3] See Nordmann, *t.* vii. viii.

Amongst insects are a variety of animals which are known to walk against gravity, we see the common flies, and other two-winged and four-winged insects, walk with ease upon the glass of our windows, and course each other over the ceilings of our apartments, without, in either case, falling from their lubricous, or seemingly perilous station. Writers on the subject are not agreed as to the means by which this is effected, some supposing that it is by atmospheric pressure, produced by suckers;[1] while others maintain that it is by a thick-set brush, composed of short bristles, on the underside of the foot, or by certain appendages at the apex of the claw joint of that organ.[2] Probably both these causes are in action, for though the pulvilli or foot-cushions of flies may adhere by mechanical means, those of some *Hymenoptera* and *Orthoptera* seem evidently furnished with suckers.[3] In both cases the design of an Intelligent Cause is apparent; His wisdom, which, under different circumstances, contrives different means to attain the same end; His power, which gives effect to that purpose and contrivance; and His goodness, which causes every varied mean to subserve to

[1] *Philos. Trans.* 1816. 322. *t.* xviii. *Introd. to Ent.* ii. 322. White's *Selborne*, ii. 274. *Ed. Markw.*

[2] Blackwall in *Linn. Trans.* xvi. 487.

[3] *Philos. Trans.* ubi sup. *t.* xix. xxi.

the more convenience and comfort of the animals in which each obtains. Could we trace exactly the history and habits of every group of animals, nay, of each individual species, we should discover that the slightest variation was to answer a particular end; and that even its very hairs and pores were all numbered with reference to special uses, foreseen by Divine Wisdom.

Amongst other purposes for which suckers were given to the Class of Insects, one bears relation to the intercourse of the sexes. This is particularly observable in the males of the predaceous beetles,[1] especially the aquatic ones. In the terrestrial ones[2] indeed something of the kind takes place, for the males may be known by having the three or four first joints sometimes only of the anterior tarsi, and sometimes of the intermediate, more or less dilated and furnished underneath with short bristles, intermixed, it should seem, with very minute suckers, and in some with transverse ones.[3] But these organs are most conspicuous in the male of our most common water-beetles,[4] in which the three first joints of the anterior tarsus form a dilated orbi-

---

[1] *Carnivora.* Lat.

[2] *Cicindelidæ, Harpalidæ, Carabidæ,* &c.

[3] E. G. *Harpalus caliginosus.* F.

[4] *Dyticus marginalis,* &c. *Philos. Trans.* ubi supr. *t.* xx.

cular shield, covered with minute suckers, sitting on a tubular foot-stalk, with two exceeding the rest greatly in size. The intermediate legs also have the three first joints thickly set with minute suckers.

Leaving the invertebrated animals the occurrence of suckers becomes very rare; very few instances are upon record, in the whole Sub-kingdom of vertebrated animals, of this kind of formation, two in the Class of fishes and the other in that of reptiles, namely the lump-fishes,[1] the sucking-fishes,[2] and the Gecko lizards.[3] Under the name of *lump-fishes* I include all those whose ventral fins unite to form a disk or sucker by which they are enabled to adhere to the rocks, constituting Cuvier's family of *Discoboles*. But the most celebrated of this tribe, in ancient as well as modern times, are the sucking-fishes or *Echenëis*, which Pliny says were so called from their impeding the course of the vessels to which they adhered. On the back of their head they have an oval cotyloid disk fitted with numerous transverse laminæ denticulated at their posterior edge, forming a double series; by the aid of this apparatus, which appears to adhere by means of the teeth of its laminæ as

[1] *Cyclopterus Lumpus*, &c.
[2] *Echenëis.*
[3] *Gecko.* Daud. *Stellio.* Schn. *Ascalabotes.* Cuv.

well as by suction, this animal attaches itself to the whale, the dolphin, the shark, the turtle, and other inhabitants of the waters, and even to vessels that are sailing, and thus organs, which at first sight appear to stop all locomotion in the animal, are the means which enable it, like certain barnacles,[1] to traverse half the globe. The fins of this animal do not permit it to swim with ease and velocity ; and therefore this must be regarded as a compensating contrivance, by which it can the more readily fulfil its functions and instincts. Though they are disengaged with difficulty by human force from the vessel to which they are fixed, they very easily detach themselves, and swimming on their back, pursue any object that attracts their attention or excites their cupidity.

It is singular to remark that in the case of two such animals, as the barnacle amongst the *Cirripedes*, which has naturally no locomotive powers and organs; and the Echenëis amongst the fishes, in which they are insufficient to transport it far from its native rocks and haunts, such means should be afforded by a kind Providence of visiting in safety the most distant oceans. These animals, though they may be called parasitic, from their adhering to other animals, yet,

[1] See above, p. 5.

as they do not appear to imbibe any nutriment from them, the design of this singular instinct seems to be merely their transport, for purposes not yet fully ascertained.

But there are other fishes whose mouth is a suctorious organ, analogous to that of the leech, by which they suck the blood of the aquatic animals they adhere to ; of this description are the *Lamprey*[1] and the *Hag*,[2] but upon these I shall not further enlarge.

The other sucker-bearing vertebrated animals, which I mentioned, were those Saurians which form the genus *Gecko*, and the object of this structure, in them, is to enable them to walk against gravity, that thus they may be empowered to pursue the insects, possessing the same faculty, up perpendicular or along prone surfaces. These suckers,[3] consisting of transverse laminæ, occupy the terminal part of the underside of the toes. By aid of these organs they can mount the smooth chunam walls of houses in India. Another Saurian genus,[4] the Gecko, of the West Indies, has a similar organ, by means of which it climbs up trees, as well as the walls of houses, in the pursuit of insects.

The adhesion of suckers and their relaxation,

---

[1] *Petromyzon.*                    [2] *Myxine.*
[3] *Philos. Trans.* 1816. *t.* xvii. *f.* 2.          [4] *Anolius.*

especially in locomotion, in order to answer the end for which they were given, must be as perfectly dependent upon the will of the animal, as our steps on the plane we are moving on are upon ours; and yet in some instances, as in the perch-pest,[1] the animal, when once fixed, can scarcely disengage itself; but in this case, having attained its ultimate station, this is of no importance.

If we study the individual cases of all the sucker-bearing animals, we shall find that this kind of organ was necessary, and all its modifications, to enable them to fulfil effectually their several instincts, and to do the work appointed them by their allwise Creator.   For instance, in vain would the Cephalopods pursue and endeavour to seize and devour the crab or the lobster, if, instead of tentacles set with numerous suckers, they had the paws and retractile claws of the Feline race: or how would the Gecko be enabled to overtake its insect provender, if its feet were like those of the rest of its class?

As supplementary to this account of suckers, I may mention a locomotive organ, given to a very numerous tribe of invertebrated animals, which, as I observed on a former occasion, appears in some degree to partake of the nature

---

[1] *Achtheres Percarum.*   See above, p. 118.

of a sucker, and which is eminently adapted to the structure, circumstances, and wants of the animals that are provided with it. I mean the expansile foot of the great majority of Molluscans : these animals are the only instance of a *unipede* structure in creation, but this one foot answers every purpose of a hand or leg ; it spins for the bivalves their byssus,[1] is used by others as an auger,[2] by others as a trowel,[3] and by others for other manipulations, and is generally their sole organ of locomotion : from its soft and flexible substance it can adapt itself to the surfaces upon which it moves, and by the slime that it copiously secretes lubricates them to facilitate its progress. In very dry weather, however, it cannot move with ease over the arid soil, but when humid from rain, the whole terrestrial Molluscan army issues forth, naked, or in various panoply, each according to its kind, covering the face of the earth, so that it is not easy to avoid crushing them.

The most careless observer of God's creatures must be struck by the correspondence between this foot, and the animal to which it is given ; had its locomotions been by means of an organ of a solid substance, or by means of several such organs, the harmony of structure which now

---

[1] Vol. I. p. 251.     [2] *Ibid.* p. 246.     [3] *Ibid.* p. 289.

strikes us, and relationship between its different parts would be done away, and we should think we beheld a mongrel monster engendered by strange mixtures of animals, rather than a creature harmoniously moulded by the hands of an allwise Creator.

I may also mention here a few other organs which seem to present some analogy to suckers, and which, though aiding in locomotion, are not, strictly speaking, locomotive organs, or those by which locomotion is effected. I allude to the spurious legs, or prolegs of the larves of insects. These are usually retractile fleshy organs, analogous to the bristle-armed protuberances of the Annelidans, rendered necessary by the length of these animals, and supporting them as props, and which usually, by means of a coronet or semicoronet of hooked spines or claws, and by applying their prone surface to the plane of position, take strong hold of it: these legs do not step; the six anterior jointed legs, where they exist, are the walking legs; but these organs having been fully described in another joint work of Mr. Spence and myself,[1] I must therefore refer the reader for further information on the subject to that work.

What are called the *pectines* or comb-like

---

[1] *Intro. to Ent.* iii. 134.

organs of scorpions, and those pedunculated ones which are attached to the hind legs of the *Solpuga* or *Galeodes*, are conjectured by M. Latreille to be connected with the respiration of these animals. Amouroux seems to regard the former as a kind of sucker, but no actual observations have as yet ascertained their real nature, except that the author last named, states that he has seen the animals use them as feet.

*Setæ* or *Bristles.* Having fully considered suckers and their analogues, I shall next advert to a species of locomotive organ, principally confined to the *Annelidans*, animals whose locomotions are chiefly produced by the contraction and expansion of the rings of which their body is composed, but which are also furnished with lateral setiform organs, which assist them in their motion, by pushing against the plane of position.

The majority of these animals are aquatic, and some of them grow to a great size; I have a specimen, which I purchased from the collection of the late lamented Mr. Guilding, which is more than a foot long, and as thick as the little finger: it has a double series of what may be denominated its legs, each furnished at its extremity with a bunch of very fine retractile bristles, and those of the dorsal series having besides a branchial organ or gill on each side, consisting

of numerous threads.  This remarkable animal
appears to belong to Savigny's genus *Pleïone*,
and is probably his *P. pedunculata*, and the
*Nerëis gigantea* of Linné.  The bristles in these
legs seem not calculated for pushing on a solid
surface, but are rather organs of natation, analogous, in some degree, to the branching legs of
the Branchiopod Entomostracans.  In the earth-
worms[1] the lateral bristles are simple, and used
to assist their motions, either on the surface, or
when they emerge from the earth, or make their
way into it.

At first sight, one would not suppose the
bristles of the Annelidans to be analogues of
jointed legs, or preparatory to their appearance
in the great plan of creation; but when we
reflect upon the approach which many of the
Nerëideans of Savigny make to the *Myriapod
Condylopes*,[2] and that these bristle-bearing legs,
in Mr. Guilding's genus *Peripatus*,[3] begin to
assume the appearance of articulations, and are
armed at their apex with claws;[4] it seems clear
that the bristles of the Annelidans, and the base
within which they are retractile, are really legs,
and lead the way to the jointed ones of the Condylopes.

I have before noticed the conversion of legs

---

[1] *Lumbricus.*
[2] See VOL. I. p. 346. PLATE VIII. FIG. 1, 4.
[3] *Ibid.* FIG. 1.          [4] *Ibid.* FIG. 2. *c. c.*

into oral organs, or their use as auxiliaries to them in the case of the Myriapods.[1] Mr. Savigny, in his description of an animal,[2] which seems the analogue of the electric centipede,[3] observes that its four anterior legs are converted into tentacular cirri, affording an additional argument for the ancient opinion that the *marine* Myriapods, as they might be denominated, have some affinity with the *terrestrial*, since, at least in this instance, the same number of legs are used as auxiliaries to the mouth.

The great majority of the Annelidans inhabit the water, and the tufts of bristles, sometimes forming fans, issuing in many cases from a dorsal and ventral conical protuberance, denominated by Savigny oars, and occasionally expanding so as somewhat to resemble them, seem in some degree analogous to the branching legs of the Branchiopod and Lernean Entomostracans,[4] and are probably natatory as well as ambulatory organs, and means by which their Creator has fitted the locomotive ones to make their way through the matted sea-weeds and the mud, when creeping after their prey, as well as to row through the water like a stately bireme. These oary feet, emulating in number those of the terrestrial Myriapods, and forming moreover,

---

[1] See above, p. 76.
[2] *Lycoris ægyptia.* Pl. VIII. Fig. 4.
[3] *Geophilus electricus.*     [4] Plate IX. Fig. 3.

as was before stated, both a dorsal and ventral series, must enable them to move with considerable rapidity: those indeed that have observed their proceedings, describe them as both swimming and running with admirable ease and speed.[1]

There is a Class of vertebrated animals, the *Ophidians* or serpents, which exhibit considerable analogy to many of the Annelidans, not only by their form and undulating movements, but also by the organs which effect their progressive motions, not indeed by means of bristles, but of parts that, pushing against the plane of position, propel the animal in any direction according to its will.

But the way in which this is effected having been clearly and most ably explained by an eminent and learned physiologist,[2] I need not here enlarge upon it, but only observe that the motion of one tribe of the Myriapods, though produced by *legs*, exactly imitates that of the Ophidians, though produced by *ribs;* and very amusing it is to see the propagation of it from one extremity to the other in the Millipedes, like wave succeeding wave in the water: a still more striking analogy, as has been already remarked,[3] is exhibited by the larger centipedes,

---

[1] See Otho Fabricius *Faun. Groendland*, 289, 298, &c.

[2] Dr. Roget.

[3] See above, p. 68.

which seem almost models of the skeleton of a serpent.

Serpents thus can move not only horizontally, but also up the trunks of trees, probably in a spiral direction, and some are said to have the power of darting from one tree to another. As these animals are not annulated, like the Annelidans, and cannot originate and continue motion by the alternate contraction and extension of the rings or segments of their body, which the nature of their integuments, their vertebral column, and muscular fibre probably preclude, the wisdom of their Creator has subjected their ribs to their will, so that they can use them as motive organs.

*Natatory Organs.*—The spurious bristle-armed legs of the Annelidans, especially those of *Peripatus*,[1] have as it were led us to the mighty host of animals furnished with *articulated* locomotive or prehensory organs, or real legs and arms, varying in number—but as these will best finish the subject, I shall first consider those external instruments of motion which are peculiar to animals inhabiting the water, or moving through the air, beginning with the first, or those distinguished by *natatory* organs. I have already mentioned some of this description, as

---

[1] PLATE VIII. FIG. 1, 2.

the oars of the paper nautilus[1] and Annelidans,[2] and also the sails expanded by the former animal and several Molluscans.[3]  Before I consider the organs in question, where they are most conspicuous, in the fishes, I must give some account of those to be met with amongst the invertebrated animals, particularly the Condylopes.  Several of the Cephalopods and Pteropods, and other Molluscans, have natatory appendages; in the former, as to many species, looking like little wings, often nearly round, attached to the lower part of the mantle that envelopes them;[4] and in the latter assuming the shape and station of the dorsal and other fins of fishes,[5] though totally different in their structure, not being divided into jointed rays as in the animals just named.

Having mentioned these, I shall next advert more fully to the organs by which the great Sub-kingdom of animals with articulated legs move in the waters, whether they always inhabit them, or occasionally visit them.  They may be divided into *three* distinct kinds.  1. Jointed legs dilated towards their extremities, as in the common whirl-wig,[6] the little beetle that forms

---

[1] See Vol. I. p. 312.          [2] See above, p. 129.

[3] See Vol. I. p. 264.          [4] Plate VII. Fig. 1.

[5] Plate V. Fig. 6, 7, 8.       [6] *Gyrinus.*

circles in the water, and in the tribe of crabs termed swimmers,[1] these I would call *Pediremes*. 2. Jointed legs, that terminate in a fasciculus of setiform branches, and are also connected with the respiration of the animal, these might be denominated *Branchiremes*, and are found in the Branchiopod Entomostracans.[2] 3. Those in which the inner side of the jointed leg has a dense fringe of hairs, called by Linné, by way of eminence, *pedes natatorii*, such as are found in many diving[3] and other aquatic beetles, these might be named *Setiremes*. As the spurious legs to which the eggs are attached, observable on the underside of the abdomen of the female lobster, cray-fish, and other long-tailed Crustaceans, are used also as natatory organs, they are ciliated for that purpose, and belong to this tribe. The same observation will also apply both to maxillary legs, and other legs of several animals of that Class. The velocity with which the diving-beetles move in the water by the action of these legs, and their suspension of themselves at the surface, by extending them so as to form a right angle with the body, when they come up for air, and the weather is fine and the water clear, affords a very interesting spectacle.

[1] *Nageurs.* Lam.   [2] PLATE IX. FIG. 4. *c.*   [3] *Dyticus.*

Amongst natatory organs I must not overlook the *tails* of the long-tailed Decapod and several other Crustaceans, which terminate in a powerful natatory organ, consisting usually of five plates, densely ciliated at their apex, the intermediate one formed of the last segment of the abdomen, and the lateral ones articulating with a common footstalk giving them separate motion, the outer consisting sometimes of two articulations, as in the common lobster, and sometimes of only one, as in the thorny lobster; the intermediate plate, as in *Galathea*, sometimes consists of two lobes; these laminæ when expanded form a most powerful natatory organ, which, if we consider the weight of their body, must be necessary to keep them from sinking, and by its vertical motion to enable them to rise or sink in the water. But natatory organs are not confined to those of the trunk and abdomen, even those of the head sometimes assist in this kind of motion. Thus in *Cypris*, an Entomostracan genus, resembling a muscle, the mandibles and first pair of maxillæ have branchial appendages used also in swimming, and their antennæ are likewise terminated by a fasciculus of threads, which, according to Jurine, the animal developes, more or less, as it wants to move faster or slower.[1]

[1] Latr. *Cours D'Ent.* i. 430.

But the most important natatory organs are those which enable the *vertebrated* inhabitants of the waters, from the giant whale to the pigmy minnow, to make their way through the waves; it will be interesting to trace the analogies of the fins of these animals to the locomotive organs, whether wings or legs of other animals, especially Mammalians. Some we shall find *sui generis*, and calculated particularly for the circumstances in which the Creator has placed the great Class of *fishes* and the rest of the marine animals; and others, in the course of our analysis, we shall observe gradually assuming the character and uses of an arm or leg.

The fins of fishes are membranes, usually supported by osseous or cartilaginous rays, which can open or shut, more or less, like a fan, but in some instances they consist of membrane without *rays*, and in others of rays without membrane. The rays are usually divided into two kinds; those which consist of a single joint, usually less flexible and pointed, whence they are called *spiny rays*, and those which consist of numerous small articulations, generally branching at their extremity, which are called *jointed rays*, these jointed rays may be regarded as precursors of the phalanxes of fingers and toes in the hands and feet of the terrestrial vertebrated animals. The first pair of fins, which are seldom wanting, and answer to the *fore-legs* or arms of those animals, are called *pectoral,* and are usually placed on the side behind the gill-covers. The second pair, supposed to be analogous to the hind-leg, are called *ventral,* and are placed under the abdomen. Besides these, there is often a fin along the back, sometimes subdivided, named the *dorsal* fin ; another under the tail, called the *anal,* and the tail itself terminates in a fin, one of the most powerful of all, which is named the *caudal,* and in some respects may also claim to be regarded as the analogue of the legs.

The, so called, fins of Cetaceans, are not properly fins, but legs adapted to their element as marine animals, the anterior pair having all the bones proper to those of mammiferous animals, covered with a thick skin, and wearing the appearance of a fin. In the sea-cow there are rudiments of nails in their pectoral fins, and they use them, both for crawling on shore, and for carrying their young, on which account they are called *Manatins,*[1] of which *Lamantins,* their French name, is probably a corruption. The tail also of the Cetaceans, which is in the shape of the caudal fin of fishes, and somewhat forked, but placed horizontally, contains some bones, which appear like rudiments of those of legs, thus, for their better motion in an element they

[1] *Manatus Americanus.*

never leave, covered by their Creator with a tendinous skin, and enabling them by an up and
down motion to sink to a prodigious depth, or to
rise from the bottom to the surface of the ocean.

If we go from the Cetaceans to the *Amphibians*,
we see a further metamorphosis of the organs of
motion. The pectoral fins of the former are now
become arms, with phalanxes of fingers, claw-
armed, but still connected by skin for natatory
purposes, and their caudal fin is converted into
rudimental legs, with a very short intervening
tail, and these legs are still of most use in the
water. These circumstances induce some suspi-
cion, especially when we consider that the caudal
fin of fishes is their most powerful locomotive
organ, that it is the real analogue of the hind-
legs of the terrestrial mammalians.

The ventral fins sometimes seem to change
place with the pectoral ones. This is the case
with the fishing-frog tribe, in which the former
are nearest to the head, and seem analogous to a
pentadactyle hand, while the pectoral ones resemble a leg and foot, and the creature looks
like a four-footed reptile.[1] The Rays,[2] in a
system, are placed at a wide distance from these,
and yet they possess several characters in common, particularly in having the hinder part of the

---

[1] See PLATE XIII. FIG. 1, 3. *Lophiadæ. Lophius.* L.
[2] *Raiadæ. Raia.* L.

body attenuated into a tail more or less slender,
and the enormous mouth and gullet of others[1]
are armed, as in the sharks, with a tremendous
apparatus of teeth.  Cuvier observes of one of
them,[2] that it can creep on the earth by means
of its fins, like small quadrupeds, and that their
pectorals discharge the function of hind-legs;[3]
so that there seems some ground for thinking
that they are a branch diverging from the Selacians towards the Reptiles.

Fins, and their analogues, were given to aquatic
animals, it should seem, solely for locomotion;
and could we witness the motions of their different
tribes, each in its place, and observe the play of
these appendages, we should find them all so
located in the body of the fish, and so nicely measured with regard to volume and weight, as to
suit exactly the wants of the animal in its station, and to act as a mutual counterpoise, so that
it should not be overswayed by the preponder-
ance of one organ over another; every thing
proving that the momentum and action of each,
both independently, and in concert with the rest,
had been nicely calculated before its creation,
by one whose Wisdom knew no bounds, whose
Will was the well being and well doing of his
creatures, each in its place, and whose Power

---

[1] PLATE VIII. FIG. 3.
[2] *Chironectes.*
[3] *Regne Anim.* ii. 251. Last Ed.

enabled him to give being to what his Wisdom planned, and his Will decreed.

Nothing is more graceful and elegant than the motions of fishes in their own pure element. Not to mention the shifting radiance of their forms, as they glance in the sunbeam; their extreme flexibility, and the ease with which they glide through the waters, gives to their motions a character of facile progress which has no parallel, unless, perhaps, in the varied flight of the wing-swift swallow, amongst their analogues, the birds. How rapidly do they glide, and are lost to our sight by a mere stroke of their tail! at another time, less alarmed, how quietly do they suspend themselves, and cease all progressive motion, so that we can discover them to be alive only by the fan-like movement of their *pectoral* fins, an action which seems, in some sort, connected with their respiration; for they move them, as I have observed, more rapidly, when, in sultry weather they seek the surface, and their muzzle emerges. These fins, the analogue, as has been before observed, of the hand or fore foot, except in a few instances, may be regarded as usually the first pair of oars that propel the vessel. Some fishes, in front of these, have another locomotive organ and weapon,[1] not intended, however, for motion so much in the *water* as on the *earth;* this is a powerful, and,

[1] PLATE XII. FIG. 1. *a.* 2.

usually, serrated bone,[1] articulating with the shoulder bones, and is to be found in the Siluridans, with the exception of the electric species, which its Creator has fitted with other arms.

The second pair of fins, as they most commonly occur, are the *ventral*, but sometimes, where fishes have a large head, they are placed forwarder, and in general they are under the most bulky part of the body ; by this arrangement, we may gather that they are intended to counteract the force of gravity, as well as to act as oars. These fins are wanting in all the fishes called, on that account, *apodes*, or footless, to which the eels, and other serpentine fishes belong, some of which also have no pectorals.

The *caudal* or tail fin, which directs the locomotions of fishes as a rudder, and gives to them the chief part of their force and velocity, in the majority of real fishes is vertical, but in flat-fish, which have no natatory vesicle, it is horizontal, as it is likewise in the Cetaceans and Amphibians ; in all these, its motion is vertical.

The *dorsal* is also a powerful fin, consisting of spiny rays ; in some tribes, as the perch, though wanting in others, it is sometimes divided into two or three fins. By its various undulations, and by the differently inclined planes which it

---

[1] N. B. The figure of the bone (2) in the Plate was taken from one dug up in this neighbourhood in forming a manure heap, which Mr. Owen informed me belonged to a *Silurus*.

presents to the water, this fin augments the means of fishes to move in any direction, and adds much to the speed with which those last named pursue their prey: it counterbalances the effect of the caudal fin in cross-currents; but, if the animals could not depress it, it might occasionally destroy the equilibrium, and overset them.

The *anal* fin seems, in many fishes, intended as an antagonist to the dorsal, to prevent the above effect and maintain the fish in its due position.

But fins were given to fishes not only to be the instruments of motion in their own element, but likewise in that of terrestrial animals; to some they were given to enable them, under particular circumstances, to vie with the birds in their aërial flights; to others, that like quadrupeds, they may undertake excursions upon *Terra firma;* and to a third description, amongst other means, to assist them in climbing the trees in quest of their food. Every body knows that the pectoral fins of the different species of flying fishes are very long; that by them, when leaping out of the water to avoid the pursuit of their enemies, the bonito,[1] and other rapacious fishes, they are supported in the air for a short time; but the action is really not *flying*, since they use these fins merely as an aëronaut, in descending,

[1] *Scomber Pelami.*

uses a parachute, for a support in the air; in fact, flying from aquatic enemies, they are soon attacked by aërial ones, and the frigate,[1] and other marine birds, make them their prey—so that they take short flights, as well as short voyages—and though they swim rapidly, they are soon tired, which is the means of saving those that escape from their numerous enemies, and preventing the extinction of the race. Besides the common flying-fish,[2] the *Pegasus*,[3] a small fish, inhabiting the Indian ocean, when pursued, leaps out of the water, and takes a short flight.

I mentioned on a former occasion,[4] the terrestrial excursions of the *Hassar*, and from the statement of Piso, in his Natural History of the Indies, published in 1658, and from that of Marcgrave, of Brazil, quoted by Linné in the *Amœnitates Academicæ*,[5] it appears that the *Callicthys*[6] migrates in the same way. Dr. Hancock mentions a fish, perhaps a *Loricaria*, which has a bony ray before the ventral as well as the pectoral fins, and which creeps on all fours upon the bed of the rivers, perhaps even when they are dry. These little quadruped fishes must cut a singular figure upon their four stilts.

---

[1] *Tachypetes Aquila.*

[2] *Exocœtus exiliens* in the Mediterranean, and *E. volitans* in the ocean, but doubts are said to rest upon this species.

[3] *P. Draco, volans,* &c.     [4] Vol. I. p. 120.

[5] I. 500. *t.* xi. *f.* 1.     [6] Plate XII. Fig. 1.

I have given a full account of a climbing fish amongst the migratory animals,[1] and shall therefore now take my leave of the finny tribes.

Perhaps the fins of the Cetaceans and Amphibians, above described, inasmuch as they are enveloped not in a membrane, like the fins of fishes, but are real feet adapted to their element, may be regarded as more analogous to what are called *paddles*, by which term the natatory apparatus of the Chelonian reptiles, and of the marine Saurians, hitherto found only in a fossil state, are distinguished. These in the former, the turtles, are formed by the legs and toes being covered by a common skin, so as to form a kind of fin, the two first toes of each leg being armed with a deciduous nail. The coriaceous turtle,[2] the parent of the Grecian lyre, which presents no small analogy to the Amphibians, has no scales either upon its body or feet, but both are covered with a leathery skin, even its shell resembling leather, and therefore it connects the paddles of the Chelonians with those of the marine Mammalians. It may be defined as a natatory organ, formed of several jointed digitations, covered by a common leathery or scaly integument. In the fossil Saurians the paddle appears to be formed of numerous bones, arranged in more than five digitations, but it is shorter and smaller, and seems better calculated

[1] Vol. I. p. 123.        [2] *Sphargis coriacea.*

for still waters and a waveless sea than to contend with the tumultuous fluctuations of the open ocean.[1]

Next to the *paddles* of the Turtles, and fossil Saurians, come the palmated or web-foot of the aquatic tortoises, and of numerous oceanic birds, in which the toes are united by a common skin. In the paddle the leg and toes together form the natatory organ; in the palmated, or lobed foot, the toes. Thus from fins we seem to have arrived at digitated legs.

*Wings.*—Turning from the denser medium of water, we must next inquire what organs have been given to animals by their Creator to enable them to traverse the rarer medium of air, to have their hold upon what to the sight appears a nonentity, and to withstand the fluctuating waves of the atmospheric sea, and the rush of the fierce winds which occasionally sweep through space over the earth. The name of *wings* has by general consent been given, not only to the feathered arm of the *bird*, but also to those filmy organs extended, and often reticulated, by bony vessels—the longitudinal ones in some degree analogous to the rays of the fins of the fishes, especially of the flying fishes—which so beautifully distinguish the *insect* races; as well as to the rib-supported membrane forming the flying

[1] See *Philos. Trans.* 1816. *t.* xvi. and 1819. *t.* xv.

organs of the *dragon;* and those hand-wings by which the *bats* with so much tact and such nice perception steer without the aid of their eyes through the shrubs, and between the branches of trees; those also of other mammi-ferous animals, such as the *flying squirrel* and *flying opossum* use in their leaps from tree to tree.

Savigny is of opinion that certain dorsal scales, in pairs, observable in two of the genera[1] of his first family of Nerëideans,[2] are analogous to the elytra and wings of insects: this he infers from characters connected with their insertion, dorsal position, substance and structure, but not with their uses and functions; for, as he also states, they are evidently a species of vesicle, communicating by a pedicle with the interior of the body, which, in the laying season, is filled with eggs,[3] a circumstance in which they agree with the egg-pouches of the Entomostracans; and therefore Baron Cuvier's opinion, that there is little foundation for the application of this term to these organs[4] seems to me correct.

Wings may be divided into organs of *flight* and organs of *suspension.* The first are found in *insects,* in which they are distinct from the legs;

---

[1] *Halithea* and *Polynoe.* See *Aphrodita Clava.* Montague in *Linn. Trans.* ix. 108, *t.* vii. *f.* 3.

[2] *Aphroditæ.*        [3] *Syst. des Annel.* 27.

[4] *Regn. Anim.* iii. 206.

in *birds*, in which the anterior leg of quadrupeds becomes a wing; and in *bats* and *vampyres*, in which both the anterior and posterior legs support the wing.

The second kind of wings is found in the *flying cat*, the *flying squirrel*, and the *flying opossum*; and, under a different form, in the *flying dragon* of modern zoologists.

The wings of *insects* differ materially from those of birds, and of certain Mammalians: for instance, the bats and vampyres, since in them they are not formed by skin or membrane, attached to the fore leg, or both legs, but are distinct organs implanted in the trunk, usually leaving the animal its classical number of legs, for its locomotions on *terra firma*. These organs are composed of two membranes, closely applied to each other, and attached to elastic nervures issuing from the trunk, and accompanied by a spiral trachea or air-vessel. These nervures vary in their number and distribution: in some insects the wing has none except that which forms its anterior margin,[1] and in others the whole wing is reticulated by them;[2] the longitudinal ones often give an inequality to the surface, and form it into folds, which probably, in flight, it can relax or contract according to cir-

---

[1] *Psilus*, &c.　See Jurine *Hymenopt. t.* v. and xiii. G. 48.
[2] *Libellulinæ.*

cumstances. In some genera[1] the wing is folded
longitudinally in repose, and in others also transversely.[2] In the higher animals the wings never
exceed a single pair; but in insects the typical
number is four; and though some are called
*Dipterous*, or two-winged, yet even a large proportion of these have, in the winglets,[3] the rudi-
ment of another pair. The anterior pair, called
elytra, &c. in the beetles, and some others, are
principally useful to cover and protect the wings
when unemployed, still they produce some effect
in flight, and they partake in a reduced degree
of the motion of the wings, those of the cock-
chaffer[4] describing an arc equal to only a fourth
part of that of the latter organs.

M. Jurine, in which he is followed by M.
Chabrier, has regarded the primary wing of insects as analogous to the wing of birds; but
though this may hold good in some respects, it
does not in its main feature. If we consider
that the wing of birds is really the analogue of
the fore-leg of quadrupeds, and replaces it; and
also that insects have a representative of that
leg fixed to the anterior segment of the trunk,
thence called the *Manitrunk*, in contradistinction
to the *Alitrunk*, which bears the wings; it seems
not probable that the anterior leg, and the anterior wing which belong to *different* segments,

[1] *Vespidæ.*　　　　[2] *Coleoptera.*
[3] *Alulæ.*　　　　[4] *Melolontha vulgaris.*

should be analogues of the same organ. The first pair of wings, or their representatives, the elytra, are connected with the hip-joint,[1] by an intermediate piece called the scapular;[2] and the posterior wings are connected with the same joint of the posterior legs by the *parapleura*,[3] so that, in some sort, the wings of insects may be regarded as appendages,—not of the *fore-legs*, or arms, which are the real analogues of the fore-leg of quadrupeds, and wing of birds,—but the first pair of the mid-legs, and the second of the hind-legs.

Some winged insects, especially the dragon-flies, like the crabs and spiders, can retrograde in their flight, and also move laterally, without turning; thus they can more readily pursue their prey, or escape from their enemies. The situation of their wings is usually so regulated in the majority with respect to their centre of gravity, as to enable them to maintain nearly a horizontal position in flight; but in some, as the stag-beetles,[4] the elytra and wings have their attachment in advance of that point, so that the head, prothorax, and mandibles do not fully counterpoise the weight of the posterior part of their body, occasioning this animal to assume a nearly vertical position when on the wing.

The apparatus and conditions of flight in birds

---

[1] *Coxa.* See *Introd. to Ent.* iii. 661.
[2] *Scapulare. Ibid.* 561.     [3] *Ibid.* 575.     [4] *Lucanus.*

and insects are very different, varying according
to the functions and structure of the animal.
In birds a longer and more acute anterior extremity distinguishes the wing, by which their
Creator enables them to pass with more ease
through the air; but in insects that extremity
is not a trenchant point that can win its own
way, but usually is very blunt, opposing either
the portion of a circle, or a very obtuse angle to
it; hence perhaps it is that the common dung-
beetle,[1] which is a short obtuse animal, "wheels
its droning flight" in a zig-zag line, like a
vessel steering against the wind, and thus it
flies, as every one knows, with great velocity as
well as noise. This also may be one reason
why insects have usually a greater volume of
wing than birds, and that a very large number.
are fitted and adorned with *four* of these organs,
which can sometimes hook to each other, by
a beautiful contrivance,[2] and so form a single
ample van to sail on the aerial waves, and bear
forward the bluff-headed vessel. The motions,
in the air, of numerous insects are an alternate
rising and falling, or a zig-zag onward flight, in
a direction up and down, as all know who have
observed the flight of a butterfly, or a kind of
hovering in the air, or a progress from flower to
flower, or backwards and forwards and every way

---

[1] *Geotrupes stercorarius,* &c.    [2] *Mon. Ap. Angl.* i. 108.

in pursuit of prey,—how admirably has their Creator furnished them to accomplish all these motions with the greatest facility and grace. And though their wings are usually naked, without any representative of those plumes which so ornament the wings of birds, and give them as it were more prise upon the air, yet in one numerous tribe,[1] the moths and butterflies, they rival the birds, and even exceed them, both in the brilliancy of the little plumes, or rather *scales*, which clothe the wings, and the variety of the pattern figured upon them, and likewise of their forms and arrangement. So that every one, who minutely examines them in this respect with an unbiassed mind, can hardly help exclaiming,— I trace the hand and pencil of an Almighty Artist, and of one whose understanding is infinite, and who is in himself the architype of all symmetry, beauty, and grace !

The wings of a variety of insects, though few, save the *Lepidoptera*, are ornamented with scales, are planted with little *bristles*, more or less numerous or dispersed; these Chabrier thinks, as well as the scales now alluded to, amongst other uses, are means of fixing the air in flight, as well as augmenting the surfaces, and points of arrest, in each wing.[2] They also strengthen the wing and add to its weight, and doubtless have

---

[1] *Lepidoptera.*      [2] *Sur le Vol des Ins.* 24.

other uses not so easily ascertained. Hair, in scripture, is denominated *power*, and probably those fluids, which we can neither weigh nor coerce, find their passage into the body of the animal, or out of it, by these little conductors ; and thus the various piligerous, plumigerous, pennigerous, and squamigerous animals, may offer points and paths not only to the air, but to more subtile fluids, either going or coming, whose influences introduced into the system, may add a momentum to all the animal forces, or, which having executed their commission and become neutralized, may thus pass off into the atmosphere.

But of all the winged animals which God has created and given it in charge to traverse the atmosphere, there is none comparable to the great and interesting Class of *birds*, which emulating the insects on one side by their diminutive size and dazzling splendours, on the other vie with some of the Mammalians in magnitude and other characters. Here we have the humming-birds of America, scarcely bigger than the humble-bee; and there the savage condour of the same country, whose outstretched wings would serve to measure the length of the giant elephant or rhinoceros. Though we cannot mount into the air ourselves, yet every one, from the peasant to the prince, that is able to follow the flight of the birds with his eyes, is delighted

with the spectacle of life that they exhibit in the
aërial regions, and we should scarcely miss the
beasts of the earth and all the creatures that are
moving in all directions and paces over its
surface, than we should the disappearance of the
birds of every wing from the atmosphere. And
therefore the prophet in his sublime description
of the desolation of Judah, makes the disappearance of the birds of heaven the most
striking feature of his picture. *I beheld the
earth*, says he, *and lo, it was without form and
void : and the heavens, and they had no light;
I beheld the mountains, and lo, they trembled, and
all the hills moved lightly. I beheld, and lo, there
was no man, and all the* birds *of the heavens were
fled.*[1]

The wing of these animals, in many cases, so
powerful to bear them on through the thin air,
and counteract the gravity of their bodies ; to
take strong hold of that element which man
cannot subdue like water, to move through himself, and so to push themselves on, often with the
swiftness of an arrow, through its rushing winds
or almost motionless breath : the wing of birds
is in fact the foreleg or arm adapted and clothed
by Supreme Intelligence, for the action it has
to maintain, and for the medium in which that
action is to take place, and consists of nearly

---

[1] *Jerem.* iv. 23—25.

the same parts as the fore-leg in Mammalians, for there is the shoulder,[1] fore-arm,[2] and the hand,[3] with the analogue of a thumb, called the wing-let,[4] and of a finger.[5] The ten *primary* quill feathers are planted in the hand, and the *secondaries*, varying in number, on the fore-arm, these quill-feathers, being very principal instruments of the wing in flight, are also named the *remiges* or rowers of the vessel. The primary feathers usually vary in length, the external ones being the longest, so as to cause the wing to terminate in a point; those that cover the shoulder are called *scapulars;* and those short ones that cover the base of the wings above and below are called *coverts.*[6] Wings usually curve somewhat inwards, are convex above and concave below, and are acted upon by very powerful muscles. Wonderful is the structure of the feathers that compose them, and each is a master-piece of the Divine Artificer. In general it is evident that each has been measured and weighed with reference to its station and function. Every separate feather resembles the bipinnate leaves of a plant; besides the obvious parts, the hollow quill, and solid stem bearded obliquely on both sides with an infinity of little plumes; each of these latter is also formed with

[1] *Humerus.*
[2] *Cubitus.*
[3] *Carpus* and *Metacarpus.*
[4] *Alula.*
[5] *Digitus.*
[6] *Tectrices.*

a rachis or mid-rib set obliquely with plumelets, resembling hairs, and exactly incumbent on the preceding one, and adhering, by their means, closely to it, thus rendering the whole feather not only very light, but, as it were, air-tight. In the goose, the mid-rib of the plumelets of the primary feathers is dilated towards the base into a kind of keel, so that each plumelet at the summit looks like a feather, and at the base like a lamina or blade.

By the use of very fine microscopes of garnet and sapphire Sir David Brewster succeeded in developing the structure of the plumelets; he discovered a singular spring consisting of a number of slender fibres laid together, which resisted the division or separation of the minute parts of the feather, and closed themselves together when their separation had been forcibly effected.[1]

If we examine the whole wing, and the disposition and connection of the feathers that compose it, we shall find that one great object of its structure is to render it impervious to the air, so that it may take most effectual hold of it, and by pushing, as it were, against it, with the wing, when the wing-stroke is downwards, to force the body forwards. A person expert in swimming or rowing, may easily get an idea

---

[1] Lit. Gazette, Oct. 11, 1834, 690.

how this is effected, by observing how the pressure of his arms and legs, or of his oars, against the denser medium, though not in the same direction, carries him, or his boat, forwards. In the case of the bird, the motion is not backwards and forwards, but upwards and downwards, which difference, perhaps, is rendered necessary by the rarer medium in which the motion takes place.

To facilitate the progress of the bird through the air, the head usually forms a trenchant point, that easily divides it and overcomes its resistance; and often to this is added a long neck, which, in the case of many sea-birds, as wild geese and ducks, is stretched to its full length in flight; while in others, where centre of gravity requires it, as in the heron,[1] bittern,[2] &c., it is bent back.

The swiftness of the flight of some birds is wonderful, being four or five times greater than that of the swiftest quadruped. Directed by an astonishing acuteness of sight, the *aquiline* tribes, when soaring in the air beyond human ken, can see a little bird or newt on the ground or on a rock, and dart upon it in an instant, like a flash of lightning, giving it no time for escape. But though some birds are of such pernicious wing, there are others of the most gigantic size, for

---

[1] *Ardea cinerea.*  [2] *A. stellaris.*

instance the *ostrich,*[1] *emu,*[2] &c. that have only
rudiments of wings, and which never fly, and for
their locomotions depend chiefly upon their legs,
to which the muscles of power are given, instead
of to the wings.

Amongst the terrestrial animals that give suck
to their young, there is a single Family which
the Creator has gifted with organs of excur-
sive flight, and these afford the only example of
the *third* kind of those organs mentioned above.
These cannot, like insects and birds, traverse the
earth upon *legs,* as well as flit through the air
upon *wings;* for the analogues of the legs of
quadrupeds, not solely of the anterior pair, as in
birds, but of both pairs, form the bony structure
by which the wing is extended and moved, and
to which it is attached.   It will be immediately
seen that I am speaking of the *bats* and *vampyres.* These animals, which form the first
Family of Cuvier's Order of Carnivorous Mammalians,[3] are denominated *Cheiroptera,* or hand-
winged, because in them the four fingers of the
hand, the thumb being left free, are very much
elongated so as to form the supports and ex-
tensors of the anterior portion of the membrane
of which the wing is formed ; while the hind leg
and the tail, in most, perform the same office for
the posterior portion of the wing : so that two

---

[1] *Struthio.*        [2] *Casuarius.*        [3] *Les Carnassiers.*

wings appear to be united to form one ample organ of flight. The membrane itself, which forms the wing, is only a continuation of the skin of the flanks: as in the wings of insects, it is double, very fine, and so thin as to be semi-transparent; it is traversed by some blood-vessels, and muscular fibres—doubtless accompanied by nerves—which when the wings are folded form little cavities placed in rows, resembling the meshes of a net. As bats are not provided with air-cells, or air in their bones, like birds, and their flight is unassisted by feathers, these wants are compensated to them by wings four or five times the length of their body. Their flight is of a different character from that of birds, resembling rather the flitting of a butterfly; when we consider that the peculiar function of bats is to keep within due limits the numbers of crepuscular and nocturnal insects, especially moths, we see how necessary it was that they should be enabled to traverse every spot frequented by the objects their instinct urges them to pursue and devour. For this purpose their wings are admirably adapted not only by their volume, but by their power of contracting them, and giving them various inflections in flight, so that their speed is regulated by the object they are pursuing.

When we further reflect that their eyes are small and deep-seated, we may conjecture that it requires extraordinary tact and delicacy of sensation in some other organs to supply this defect in its sight. Spallanzani found that blind bats fly as well as those that have eyes; that they avoided most expertly threads of fine silk which he had so stretched as just to leave room for them to pass between them; that they contracted, at will, their wings, if the threads were near, so as to avoid touching them; as well as when they passed between the branches of trees; and also that they could suspend themselves in dark places, such as vaults, to the prominent angles. He deprived the same individuals of other organs of sensation, but they were equally adroit in their flight, so that he concluded they must have some sensiferous organs different from those of other animals to enable them to thread the labyrinths through which they ordinarily pass.

Dr. Grant observes on this subject—" Bats are nocturnal, but, contrary to what is generally the case with nocturnal animals, their eyes are minute and feeble, and indeed, comparatively speaking, of minor importance, for so exquisite is the sense of feeling diffused over the surface of their membranous wings, that they are able to feel any vibration of air however imperceptible by us; they can tell, by the slight rebound of the air, whether they are flying near any wall, or opposing body, or in free space though their

eyes be sealed or removed."[1]    A similar observation was long ago made by Mr. Bingley.[2]

We see in the circumstances here detailed a
remarkable instance of the Power, Wisdom, and
Goodness of the Creator, in compensating for the
absence or imperfection of one or more senses,
by adding to the intensity of another, and in
establishing its principal seat in organs so nicely
adapted to derive most profit by the information
communicated.

An animal nearly related to the vampyres, the
cat-ape,[3] commonly called the flying cat, and
by some the flying dog, though nearly related to
the bats, and included by Cuvier in the same
Family, differs essentially from them, in being
furnished with organs formed by the skin of the
flanks connected with the legs of each extremity,
which are calculated for *suspension* rather than
flight, being used, as Cuvier remarks, merely as
a parachute, and thus belong to the *second* kind
of wings, mentioned above.   This animal, which
climbs like a cat, vaults from one tree to another,
by the aid of the above skin, which supports it
in the air.   The *petaurists*,[4] or flying squirrels,
and the *phalangists*,[5] or flying opossums are
similarly equipped, and for a similar purpose.

---

[1] Quoted in *Lit. Gaz.* Feb. 9, 1834.

[2] *Mem. of Brit. Quad.* 34.          [3] *Galcopithecus.*

[4] *Petaurus.*          [5] *Phalangista.*

The common *squirrel*,[1] using its tail as a rudder, leaps with great agility from tree to tree, without the aid of this kind of parachute, the force of its spring being sufficient to counteract that of gravity. Providence has evidently added an organ of suspension, in the case of the three former animals, either because their vaults were necessarily longer, or because the greater weight of their bodies required it.

The dreaded name of *dragon*, attached to the monsters of fable, has excited in our imagination ideas of beings clothed with unwonted terrors, from our earliest years, so that when we find the only animal that inherits their name is an insignificant *lizard*, not more than eight inches long, we are tempted to exclaim, *Parturiunt montes*. This little animal, under the name of wings, is furnished with two dorsal appendages independent of the legs, formed of the skin, and actually supported by the six first short ribs, which, instead of taking their usual curvature, are extended in a right line. These organs are not used to fly with, but to support the animal in its leaps from branch to branch, and from tree to tree.

We see in this instance, how exactly the means are adapted to the end proposed. This animal walks with difficulty, and consequently

---

[1] *Sciurus vulgaris.*

seldom descends from the trees. It is therefore enabled to move from one part of a tree to
another, not by its legs, but by an organ formed
out of its *ribs!* How various and singular, in
this instance, as well as in that of serpents,
before alluded to,[1] are the means adopted by a
Being, who is never at a loss to answer the foreseen call of circumstances by wise expedients.

*Steering Organs.*[2]—But wings are not the only
organs of flight with which the Creator has fitted
those animals, to which he has assigned the air
as the theatre of their most striking and interesting locomotions. They would be like a ship at
sea without a *rudder*, and be altogether at the
mercy of every wind of heaven, had they no
means to enable them to steer their vessel through
the fluctuations of the viewless element assigned
to them. The eagle and the vulture would be
gifted in vain with the faculty of seeing objects
at a great distance, had they no other organ
than their sail-broad vans to direct them in their
flight. The same remark will apply as well to
the insect as to the bird, which would in vain
endeavour to discharge its functions, unless it
could steer its course according to the direction
of its will and the information furnished by its
senses. But, upon examination, we shall find

---

[1] See above, p. 130.     [2] *Gubernacula.*

that God hath not left himself without witness in this department, but hath furnished every bird and insect with such an organ of steerage as the case of each required; nay, even amongst the beasts and the reptiles we may discover similar means of directing their motions, especially when they leap, whether from the ground, or from tree to tree.

The *caudal fin*, or tail of fishes, may be regarded as belonging in some degree to this head; but as this is also their principal organ of locomotion, I thought it best to consider it with the other fins.

The *abdomen* of many insects seems to serve them as a rudder, being composed of several inosculating rings formed each of a dorsal and ventral segment; it is capable of considerable flexion in almost all directions; it can be elevated or depressed, and turned to either side, so that it seems, in a great degree, calculated to enable insects to change the course of their flight according to their will. But besides this important organ—which by the air it is constantly inspiring adds force also to the internal impulse, and to the air-vessels in the wings—insects have other auxiliaries to keep them in their right course. Whoever has seen any grasshopper take flight, or leap from the ground, will find that they stretch out their hind legs, and, like certain birds, use them as a rudder. The tails also of the day-flies[1] seem to be used by them as a kind of balancer in their choral dances up and down in the sun's declining beam.

But the most interesting and beautiful organ for steering animals in the air, is that formed by the tail feathers of birds, called by ornithologists, *rectrices*, or *governing* feathers, because they are used to direct their course; these are feathers planted in the rump,[2] usually twelve in number—but in some amounting to nearly twenty—constituting two sets of feathers of six each, and forming together a kind of fork like the caudal fin of some fishes; the inside of each feather is set with much larger plumelets than the outside, so that there is a double series of corresponding feathers beginning one on the right side, and the other on the left; the middle feathers in each series differ sometimes from the five exterior ones, being more acute, and wearing a different aspect. In flight the tail-feathers appear to be expanded, and probably the bird, by giving an impulse to either series, can turn this way or that; or by their depression or elevation, judging from their analogy with the caudal fin of fishes, rise or fall. The rudder-tail here described is that of the male bull-finch;[3] in many birds of the Gallinaceous Order, as the common cock and peacock, these feathers form a glorious

---

[1] *Ephemera.*     [2] *Uropygium.*     [3] *Loxia pyrrhula.*

ornament, but seem to lose their use as a steering apparatus. In the black game[1] the two sets of feathers of the tail turn outwards, one on each side, and so form a fork; and, in our domestic poultry, these sets of feathers, when not expanded, fold upon each other. Some of the waders,[2] the tail-feathers of which are short, use their long *legs*, like the grasshoppers, as a rudder in flight, stretched out straight behind them.

Many of the web-footed birds,[3] as the goose and duck tribes, also have these feathers very short, which seems a convenient provision for aquatic birds, but whether their legs assist in directing their course seems not to have been ascertained. Some of them, however, as the pin-tail duck,[4] have the middle feathers of the tail elongated, as they are in many other birds; in the swallow tribe,[5] and the sea-swallow,[6] the external feathers of the tail are elongated, as these birds are frequently turning when in the air, and flying backwards and forwards; their Creator has thus equipped them for their ever changing evolutions. Some birds, as the thrushes,[7] magpies,[8] and other crows, have all the tail feathers long, which gives great power to them in flight.

[1] *Tetrao Tetrix.*  [2] *Grallatores.*  [3] *Palmipedes.*
[4] *Anas acuta.*  [5] *Hirundo.*  [6] *Sterna.*
[7] *Turdus.*  [8] *Corvus Pica.*

The tails of quadrupeds, both oviparous and viviparous, appear, in many cases, to act in some degree as a rudder. They are not only useful to those lately mentioned, that by the assistance of a kind of parachute, leap from tree to tree; but likewise to the feline race, when they spring upon their prey; the tail is then extended stiffly in a right line, as if to guide them through the air straight to the object they have been watching from their lair. The long tail also of many lizards may, in their sinuous windings, serve some purpose connected with their locomotion related to the one under discussion, though we have not data sufficient to speak positively on the subject.

*Legs.*—We are now arrived at organs that are the most perfect instruments of locomotion and prehension, organs which are found in their greatest perfection in the highest animals, articulated *legs* and *arms*, terminating in the most perfect instrument, upon the due employment or misemployment of which the weal or woe of the whole human race, as far as second causes are concerned, depend.

The legs of animals may be considered generally as to their *number, composition,* and *adaptation to their functions.*

As to their *number*, taking the legs of vertebrated animals, which may be regarded, being the most perfect, as a standard to measure others by, we may assume that *four* is the most perfect number.    Thus, in man, the highest animal, there are two for locomotion, and two principally for prehension.    Taking, therefore, man for the ultimate point to which all tend, let us see how, in this respect, the scale is formed.

We observed in certain tribes of the Annelidans, an approach to jointed legs, and it should seem a link, connecting, in some degree, that Class with the *Myriapods;* with these last, therefore, we may start in our consideration of articulated locomotive organs, and here we find a long body moved by numerous legs, gradually acquired, as we have seen, with its increasing length.    We may observe, that in the superior tribes of animals, the four legs being planted in pairs at each extremity of the body, the gradual increase of stature did not require additional props, but only the proportionate growth of the existing or natal legs and arms ; but in the Myriapods, where the great increase of the body in length is not between the original extremities, but beyond them, additional supports were requisite, so that as the body increased in length, its Creator, in his goodness willed—that it might not draw its slow length along like a wounded snake —that it should be furnished at the same time with a proportionate increase in the number of its locomotive organs.    These animals then, with respect to number of legs, may be regarded as at the foot of the scale, and are the furthest removed from man.

From the Myriapods, we go to the great *Crustacean* host, in which, including the maxillary legs, the real analogue of the legs of Hexapods, the typical number is *sixteen;* and from these, the transition is naturally to the *spiders,* which have half that number, and from them to the *insect* tribes, walking only upon six legs. Having arrived at a hexapod type, we may observe that one pair of the legs has a direction towards the head, and are located in the anterior segment of the trunk; and that the other two pairs have a direction the contrary way, towards the abdomen, and are located in that part of the trunk which bears the wings, and of these, the last pair may be regarded as the representatives of the legs in man, and of the hind legs of quadrupeds.

As to the *composition* of legs, if we take the arm and leg of man for the type or standard with which to compare all the articulated organs of locomotion and prehension with which animals are gifted, we shall find a considerable, though not an entire, correspondence between them. Anatomists usually divide the *arm,* or anterior extremity, into *four* principal portions, namely, the *shoulder-blade,*[1] the *shoulder,*[2] the

---

[1] *Scapula.*  [2] *Humerus.*

*fore-arm*,[1] and the *hand*;[2] but the *leg* only into
*three*—the *thigh*,[3] the *shank*,[4] and the *foot*.[5]
The first of these, however, the thigh, inoscu-
lates with the lower part of a bone, called the
*nameless bone*,[6] which in very young subjects
forms three, named the *haunch*,[7] the *share-
bone*,[8] and the *hip-bone* :[9] now this bone appears evidently the analogue of the shoulder-
blade in the anterior leg or arm, and thus, admitting this, both extremities in the number of
principal parts correspond with each other.

As the vertebrated animals, for the most part,
agree with their prototype in the greater articulations of their anterior and posterior extremities, though much modified in particular instances and for particular uses, I shall now only
compare the legs of the great sub-kingdom of
Condylopes, or invertebrated animals with jointed
legs, with those of man, and other Mammalians,
and inquire how, in the above respect, they consist of analogous parts.

The remarkable distinction which separates
the vertebrated from the invertebrated animals,
namely, that, in the former, the muscles have no
*external* points of attachment; and, in the latter,

---

[1] *Cubitus*, including two parallel bones, the *Ulna* and *Radius*.
[2] *Manus.*          [3] *Femur.*
[4] *Crus*, including also two parallel bones, *Tibia* and *Fibula*.
[5] *Pes.*        [6] *Os innominatum.*      [7] *Os ilium.*
[8] *Os pubis.*     [9] *Os ischium.*

with a few partial exceptions, no *internal* ones—
must produce a marked difference in all parts of
their several structures, and, amongst the rest,
between their organs of locomotion and prehension: and therefore it is not to be expected that
they will be perfectly analogous in their composition. Thus in the *invertebrates* the parts corresponding with the fore-arm and shank of the
*vertebrates* do not consist of two parallel bones;
the hand and the foot also are essentially different; and the parts by which the extremities
in one case articulate with the vertebral column
towards its summit and base, and in the other
with the trunk of the animal at various points,
are usually extremely dissimilar: in several
beetles, however, the basilar joints, especially of
the hind legs, assume something of the character
and form of the shoulder-blade of Mammalians;
and in certain water-beetles[1] the posterior pair
are immoveable. In quadrupeds, usually, the
thighs are remarkably clothed with muscle,
especially towards their base; but, in the Condylopes, with the exception of some beetles and
jumping insects, where a powerful muscular
apparatus was requisite, they are not conspicuously incrassated, so as to contain muscles of
great volume.

From these circumstances I am induced to

[1] *Dytiscus.* L.

confine my observations to the *numerical* composition of the locomotive and prehensory organs
of Condylopes, and animals that give suck to
their young.

In order to perceive clearly how far they agree
or disagree in this respect, it will be adviseable
first to inquire whether these organs in Condylopes themselves can be reduced to a common
type.

The Crustaceans and Arachnidans, including
under the latter denomination all regarded by
Latreille as belonging to the Class, at the first
inspection of the organs in question, appear to
have one joint more than insects. This supernumerary joint is the *fourth*, in *The Introduction
to Entomology* named the *Epicnemis*,[1] which is
there regarded as an accessory of the shank.
But from further observation, and from a comparison of this joint of the Arachnidans with an
analogous one in the Crustaceans, in which it is
longer and more conspicuous, I feel convinced
that, short as it is in them, it is really the *shank*
in that Class, and that the long joint usually regarded as the shank is analogous to the first,
often dilated and elongated,[2] joint of the tarsus
in insects. That this joint belongs to the tarsus
or foot will be further evident from the following
circumstance. If we examine the anterior leg, or

---

[1] Vol. iii. 668.

[2] E. G. In the Bees and many other *Hymenoptera*.

arm, of the *lobster* or *crab*, we shall find that the joint in question, which is the fifth of the leg,[1] is what is called the metacarpal joint, a process of which forms the index or finger of the didactyle hand or forceps of these animals, and the succeeding and terminal joint the opposing thumb. It is evident, therefore, that this joint belongs not to the shank or cubit, but to the foot; and that consequently a Crustacean or Arachnidan leg or arm numerically corresponds in its greater articulations with that of an insect.

Having proved, I hope, to the satisfaction of the reader, that the legs of Condylopes, with regard to the number of their principal articulations are reducible to one type,—unless we may except some of the *Acaridans*, or mites, and the *Branchiopod Entomostracans*, which appear reducible to no general rule—I shall next endeavour to show that the Condylope leg does not usually differ numerically from that of the quadruped or mammalian; and that the former consists of only *four* principal articulations as well as the latter, and it will not require many words, or any laboured disquisition, to prove this. The, so called, *trochanter* is, with great propriety, considered by M. Latreille as being a joint of the thigh, as it really is, and in many cases, especially in Coleopterous insects, has no separate

[1] PLATE X. FIG. 1.

motion; consequently if this opinion be admitted, the number of articulations, both in the Condylopes and Mammalians, will be the same.

Animals that are built upon a skeleton, or incased in an external crust or rigid integument, in order to have the power of free locomotion and prehension, must necessarily be fitted with *jointed* organs, whose articulations are more numerous at the extremity, where the principal action is, that those parts may so apply to surfaces as to enable the animal to take sufficient hold of them for either of the above purposes.

There is a circumstance connected with the legs of insects which, at first sight, seems to throw some doubt upon this conclusion. The shank has often at its apex, and sometimes the cubit, certain little moveable organs, which have been called *spurs*, but which really appear to aid the animal in its locomotions,[1] and in some they even terminate in suckers:[2] as these organs are co-ordinate with the jointed tarsus, they seem in some sort a kind of auxiliary digitation. In the *mole-cricket*[3] the structure is still more anomalous, the cubit terminating in four strong digitations or claws, opposed to which is the, so called, tarsus, which seems analogous in some sort to a jointed thumb, so that the whole represents a pentadactyle hand. A

[1] *Introd. to Ent.* iii. 674.
[2] *Philos. Trans.* 1816, *t.* xix. *f.* 8. 9.        [3] *Gryllotalpa.*

similar anomaly distinguishes the posterior pair
of legs of one of the Entomostracans, the *king-
crab*: in these, besides the tarsus armed with
two claws, there are four moveable digitations.[1]

Though the Creator has evidently connected
the sphere of animals by some organs or characters common to the whole, and generally
speaking, in the tribes that we are comparing,
has formed the organs which I am considering,
as to their articulations, upon a common type;
yet occasionally we see departures from a strict
adherence to the likeness, as in the cases here
specified, where the circumstances and functions
of an animal required such departure.

*Adaptation of Legs.*—It is by the adaptation
of its legs to the circumstances of an animal, and
to the functions which it was created to exercise,
that the design of an Intelligent Cause is apparent, and the power, wisdom, and goodness of
the Creator manifested.

The well known adage, *Natura non facit
saltus*, is exemplified in the passage, with respect
to their locomotive organs, from the expansile
Annelidans to the rigid Condylopes; for in numberless instances, we have in the larvæ of insects
a kind of intermediate animal, in some degree
expansile, some of which move like the leech,[2]
and others are apodes, like worms, moving by

---

[1] Savigny, *Anim. sans Vertebr.* i. *t.* viii. *f.* 1. k.
[2] The Geometric caterpillars or loopers.

the contortions of their bodies, a large proportion
at the same time having the jointed legs of their
Class when arrived at perfection, and in their
spurious legs imitating, in some sort, the locomotive organs of the Annelidans.

The principal offices of legs are to enable the
animal to procure the kind of food which its
nature requires; to be employed in operations
connected with the continuation of its kind;
and to be instrumental in its escape from danger
and from the pursuit of its enemies; and the
means by which these ends are accomplished
are the comparative *length* of its legs; their
*volume*, either in whole or in part; the structure of their *extremity*, either for locomotion
or prehension; or where the extremity of the
legs is not adapted to the latter function, certain
compensating contrivances calculated to supply
that want.

To enable some animals to come at their food,
sometimes a great difference, as to *measure*,
between their anterior and posterior extremities,
is necessary. At the first blush, and before we
were acquainted with its habits, should we
chance to meet with a *giraffe*,[1] so striking is the
seeming disproportion of many of its parts, that
we should be tempted to take it for an abortion
in which the posterior parts were not fully de-

---

[1] *Camelopardalis Giraffa.*

veloped. Observing its length of neck and elevated withers, the apparently unnatural declivity of its back, and the comparative lowness of its hind quarters, we should conclude that such must be the case. But if we proceeded to inquire into the nature of its food, and were told that it subsisted by cropping the branches of certain trees which thus it was enabled to reach, the truth would flash upon us, we should immediately perceive the correspondence between its structure and its food, and acknowledge the design and contrivance of a benevolent Creator in this formation.

A similar idea would perhaps occur to us the first time we saw a *jerboa*,[1] or a *kanguroo*.[2] Hasselquist says of the former—that it might be described as having the head of a hare, the whiskers of a squirrel, the snout of a hog, the body, ears, and fore-legs of a mouse, hind-legs like those of a bird, with the tail of a lion ; and an ancient zoologist would have made a monster of it that might have rivalled the chimæra. The kanguroo also would have met with a similar fate. Though the jerboa is not a marsupian animal like the kanguroo, yet they have many characters in common. They both have very slender fore-quarters, and short and slender fore-legs; their hind-quarters, on the contrary, are remarkably robust and incrassated,

---

[1] *Dipus.*     [2] *Macropus.*

and they sit erect, resting upon them like a hare ; both have a long powerful tail, which they use as a fifth leg. The object of this formation, at the first glance, so at variance with all ideas of symmetry, appears to be a swifter change of place, and more ready escape from annoyance or violence. The jerboa is stated to take very long leaps, and those of the *kanguroo* are said to extend from twenty to twenty-eight feet, and they rise to an elevation of from six to eight feet. When they leap they keep their short fore-leg pressed close to their breast, and their long and robust tail, having first assisted them in their leap, is extended in a right line. A double end is answered by their peculiar structure ; sitting on their haunches, they can leisurely look around them, and if they spy any cause of alarm make off by the means just stated. Their attenuated fore-quarters and short fore-legs rendering it much more easy for them, overstepping every obstacle, to dart into the air ; their centre of gravity is then removed nearer the hind quarters, so that the tail can act as a counterpoise to the anterior part of the body.

The *jerboa* also, like the kanguroo, when alarmed, springs into the air. When ready to take flight, it stands, as it were, on tip-toe, supporting itself by its tail. Its fore-legs are then applied so closely to the breast as to be invisible, whence the ancients called it *Dipus*, or biped;[1] having taken their spring they alight upon their fore-feet, and elevating themselves again, they are off so rapidly, that they seem to be always, so to speak, upon the wing. They use their long tail to support themselves when they recover from their leaps, giving it the curvature of the letter S reversed, thus, ω. When their tail has been shortened at different lengths, it has been found that their leap is diminished in the same proportion; and when it was wholly cut off they could not leap at all.

We see, in one Order of the *Birds*,[2] the *Waders*, a remarkable disproportion of the legs to those of the rest of the Class; they look as if they walked upon stilts, whence the name of the Order, so disproportionally long are their legs to those of the generality of birds. I have before noticed the use of these legs to them in flying,[3] but the principal object of this structure is to enable them to prey upon aquatic animals, fishes, worms, and the like. Whoever is in the habit of frequenting estuaries, and other waters, will generally see some of these birds, as herons and bitterns, standing in them, where

---

[1] Herodot. *Melpom.* § 192. Ed. Reizii.

[2] It is to be observed in general, with respect to the Class of *Birds*, that the conspicuous part of their legs is not the *shank*, which is chiefly covered by muscle and feathers, but is formed of the tarsal and metatarsal bones united into one.

[3] See above, p. 164.

shallow, and ever and anon dipping their heads,
and then emerging swallow their capture.  The
design of this structure must be obvious to every
eye, namely, to qualify these birds of prey to
assist in keeping within due limits the population of the various waters of our globe, which
other predaceous animals cannot come at.

Another tribe of long legged birds, which
Cuvier considers as belonging to the present
Order, though their habits and *habitat* are altogether different, and which constitute his
family of short-winged waders,[1] is that to which
the Ostrich[2] and Emu[3] belong, but in these the
object of this structure is to fit them not for
standing in the water, but for running in the
sandy desert ; and such is the velocity of the
ostrich that it can outstrip the fleetest Arabian
courser when pursued.

Other birds are remarkable for the *shortness*
and strength of their legs ; of this description
are the *aquiline* race, which are thus fitted by
their Creator for seizing and holding fast any
prey which their piercing sight discovers.

There is one, and a very elegant bird, belonging to this Order, the secretary-bird,[4] the legs
of which are so long, that many ornithologists
have arranged it with the waders.  It is, how-

---

[1] *Echassiers brevipennes.*       [2] *Struthio Camelus.*
[3] *Casuarius Emeu.*             [4] *Ophiotheres cristatus.* Veill.

ever, very properly placed amongst the predaceous birds. Its long legs are given it to enable it to pursue the serpents, which form its food. We see, in this instance, a departure from one of the typical characters of its own tribe, and those of another adopted in order to accommodate the animal to the circumstances in which it was the Divine will to place it, and to fit it for the function which it was there to exercise.

Amongst the *Reptiles* there is little diversity, as to the relative proportions of the organs we are considering, and their parts; in the *Batrachians*, or frogs and toads, which are mostly leaping and swimming animals, the hind legs are elongated to accommodate them to those kinds of locomotion; and in some of the Saurians or lizards, which are approaching to the Ophidians or serpents, the legs are very short,[1] and sometimes reduced to a single pair;[2] even in some serpents rudiments of a pair of legs have been discovered, particularly in the *Boa*.[3]

Some *insects* are remarkable for the vast length of their anterior pair of legs; what may be the object of this formation has not been discovered except that, in one instance,[4] it is found only in one sex. The animal I allude to belongs to the tribe of Capricorn beetles,[5] and

---

[1] E. G. in *Seps*.     [2] As in *Bipes*.     [3] *Zool. Journ*. iii. 253.
[4] *Acrocinus longimanus*.     [5] *Cerambyx*. L.

seems not to be uncommon in Brazil. The fore legs of the male are more than twice the length of the body, while those of the female, though longer than the others, are scarcely half so long.

Many insects are formed, in some degree, after the pattern of the kanguroo and the jerboa, in order to enable them to transport themselves by leaping beyond the reach of their enemies. The thighs of their hind legs are incrassated so as to afford a box capable of containing muscles sufficiently powerful, by their action, to send them through the air to an almost incredible distance. If we examine the structure of the posterior legs of any common *grasshopper*, we immediately see, both from the position of the joints with respect to each other, and the shape and volume of the elongated thigh, that they are made for leaping. The shank, when the animal prepares to leap, forms an acute angle with the thigh, so that being suddenly unbent, it springs forward, often to the distance of two hundred times its own length. Many carriages are set upon springs made to imitate the position of this insect preparing to leap, which are known by the name of grasshopper springs.[1]

Several *beetles* rival the grasshoppers in their leaps, and have their posterior thighs much

[1] See *Introduction to Entomology*, ii. 310.

disproportioned to the bulk of their bodies, which allow space for a sufficient muscular apparatus, to send them, like an arrow from a bow, to a great distance. If a finger be held to a leaf covered by the *turnip flea*,[1] in the twinkling of an eye, all skip off and vanish. We may hence imagine with what expedition they disappear at the approach of any insectivorous bird. Thus their Creator, who cares for the meanest of his creatures, has furnished them with means of escape, to prevent their annihilation, and to preserve them in such force, as may best answer his end in creating them.

But besides *partial* modifications of the structure of these organs for particular uses, others are more *general* and affect the whole leg. Every one is aware how well adapted, by their fleetness, some of the *Ruminant* Mammalians are to make their escape from their ravenous pursuers, the most adroit and the most ruthless of which is the *mighty hunter*, man.

If we look at the legs and hoofs of the *deer* tribe,[2] the former long, slender, and elastic ; and the latter calculated for sure footing; and if we consider besides the quickness of their senses of seeing and hearing, we see at once that their structure is the effect of *design*, and that the

---

[1] *Haltica oleracea, Nemorum, &c.*          [2] *Cervus.* L.

deepest intellect presided at its first fabrication.[1]
Though man, as well as every ferocious beast,
pursues these beautiful and elastic animals, it is
only because he is *Gulæ deditus*, seldom with any
view to seek their alliance, or to turn them to
his purposes.   There are some, however, as well
as the rein-deer,[2] cherished by the Laplander
as his principal treasure, but pursued by the
American savage only to be devoured, which
probably might be employed with advantage, as
well as the dog, in countries not suited to our
beasts of burthen; and it has been supposed
that the Wapiti[3] might be trained and rendered
useful, I am ignorant, however, whether any
steps have ever been taken to ascertain this.

But the legs, as well as instruments of flight
and escape, are adapted in fiercer animals to
the pursuit and prehension of their prey, and in
this, and many other respects, their *hand* or
*foot* is the part principally interesting.   This is
used for so many various purposes, that perhaps
it will be best to take a summary survey, in this
respect, of all the Classes of animals with articulated legs, and briefly point at their different
structures and their uses.

As I have already given an account of the
two kinds of forceps of Crustaceans,[4] I shall

---

[1] See Roget, *B. T.* i. 506.      [2] *Cervus Tarandus.*
[3] *C. Stongyloceras.*            [4] See above, p. 37, 38.

begin with the legs of the *Arachnidans*, or
spiders. Every one who examines the web of a
common spider, whether it is formed of con-
centric circles supported by diverging rays, or
whether it imitates any finely woven substance,
will be convinced that she must be furnished
with a peculiar set of organs to effect these
purposes: that she must have something like
a *hand* to work with. Amongst the small things
that are wise upon earth, Solomon mentions
the spider; and the way by which he tells
us she shews her wisdom is by her prehensory powers—*she takes hold with her hands*.[1]
And truly what Arachne does with her hands
and her spinning organs is very wonderful, as I
shall have occasion hereafter to shew; I shall
now only make a few observations upon the
organs by which she takes hold.

Spiders are gifted with the faculty of walking
against gravity, even upon glass, and in a prone
position. According to the observations of Mr.
Blackwall, this is not effected by producing
atmospheric pressure by the adhesion of suckers,
but by a brush formed of " slender bristles
fringed on each side with exceeding fine hairs
gradually diminishing in length as they approach its extremity, where they occur in such
profusion as to form a thick brush on its inferior

---

[1] *Prov.* xxx. 28.

surface."[1]　These brushes he first discovered on a living specimen of the *bird-spider*,[2] and the same structure, as far as his researches were carried, he found in those spiders which can walk against gravity and up glass. This is one of the modes by which they take hold with their hands, and thus they ascend walls, and set their snares in the palace as well as the cottage. Whoever examines the underside of the last joint or digit of the foot of this animal with a common pocket-lens, will see that it is clothed with a very thick brush, the hairs of which, under a more powerful magnifier, appear somewhat hooked at the apex; in some species this brush is divided longitudinally, so as to form two.

But the organs that are more particularly connected with the weaving and structure of the snares of the spiders are most worthy of attention. Setting aside the hunters,[3] and others that weave no snares to entrap their prey, I shall consider those I intend to notice, under the usual names of *weavers*[4] and *retiaries*.[5]

Before Mr. Blackwall turned his attention to the proceedings of these ingenious and industrious animals, it had not been ascertained, in what respect their modes of spinning their

---

[1] Blackwall in *Linn. Trans.* xvi. 481. *t.* xxxi. *f. 5.*
[2] *Mygale avicularia.*　　　　[3] *Araneæ. venatoriæ.*
[4] *A. textoriæ.*　　　　[5] *A. retiariæ.*

webs, and the organs by which they formed
their respective manufactures differed. But Mr.
Blackwall, whose observations were principally
made upon one of the weavers[1] which frequents
the holes and cavities of walls, and similar
places, observes that it spins a kind of web of
different kinds of silk, the surface of which has
a flocky appearance, from the web being as it
were ravelled.

This web is produced, he observes, by a
double series of spines, opposed to each other, and
planted on a prominent ridge of the upper-side
of the metatarsal joint, or that usually regarded
as the first joint, of the foot of the posterior legs
on the side next the abdomen. These spines
are employed by the animal as a carding apparatus, the low series combing, as it were, or
extracting, the ravelled web from the spinneret,[2]
and the upper series, by the insertion of its
spines between those of the other, disengaging
the web from them.[3] By this curious operation,
which it is not easy to describe clearly, the
adhesive part of the snare is formed, thus large
flies are easily caught and detained, which the
animal, emerging from its concealment, soon
despa ches and devours.

The organs by which the *retiary* spiders form
their curious geometric snares have generally

[1] *Clubiona atrox.*   [2] *Mammulæ.*
[3] Blackwall, *ubi sup.* 473.

been described as three claws, the two uppermost armed with parallel teeth like a comb, and the lower one simple and often depressed; but Mr. Blackwall found, in a species related to the common garden spider,[1] *eight* claws, seven of which had their lower side toothed.[2] The object of this complex apparatus of claws simple and pectinated, is to enable these animals to take hold of any thread; to guide it; to pull it; to draw it out; to ascertain the nature of any thing ensnared, whether it be animate or inanimate; and to suspend itself. In fact the Creator has made their claws not only hands but eyes to these animals.

Besides these organs, scattered moveable spines or spurs are observable upon the legs, especially the *three* last joints, which I consider as forming the foot, but sometimes also upon the thighs of spiders, which, as they can be elevated and depressed at the will of the animal, probably are used as a kind of finger, when occasions require it.

In the multiform apparatus of these ingenious animals, as far as we understand its use, we see how they are fitted for their office, by contributing to deliver mankind from a plague of flies, which would otherwise, like those which

[1] *Epeira Diadema.* The species examined by Mr. B. was *E. apoclisa.*
[2] Blackwall, *ubi sup.* 476.

swarmed in Egypt, annoy us beyond toleration, and corrupt our land.

If *the spider taketh hold with her hands*, and spreads her snare in kings' palaces, what shall we say of the *bee*, who with her hands erects *herself* her many-storied palaces, each story consisting of innumerable chambers, far more durable, and built of a material infinitely exceeding the flimsy webs of Arachne. Her Creator hath instructed her, and fitted her with the means, to gather from every flower that blows a pure and sweet nectar, from which, received into her stomach, she elaborates the beautiful and important product of which her wondrous structures are formed; and from the same source she is also instructed to load herself with a fine ambrosial dust, which, kneaded by her into a paste, constitutes the chief subsistence of herself and the young of the community to which she belongs.

Almost every organ, implanted in her frame by her beneficent Creator, is employed by this symbol and exemplar of virtuous industry as a hand in her several works and manipulations. Her *antennæ*, those still mysterious organs, inform her in what flowers she may find honey, and which to pass by; they plan and measure her work, and by them she examines whether all is right; she also uses them to converse with her associates, and for various other purposes;

her *tongue* is likewise an instrument equally useful to her; it can assume various shapes as occasions demand; it collects the honey from the nectar-organs of the flower; it tempers the wax for building and prepares it for the action of the *mandibles*. With these last organs she works up the wax till it is fit for use. The plumy *hairs* of her body, especially in the humble-bees, are useful in detaining the dust of the anthers. Her *legs*, more particularly the posterior pair, though not used immediately in her structures, are extremely important organs, both for preparing her food and the material with which she builds her palace. At the junction of the shank, with the first joint of the foot of this pair, a kind of *forceps* is formed, by the angle at the apex of the former and the base of the latter, with which the bee takes a plate of wax from the wax-pockets under her abdomen, and delivers it to the anterior pair of legs, by which it is submitted to the action of the mandibles. The *shanks* of the posterior legs likewise on their upper side have a cavity surrounded with hairs, which form a kind of basket, in which the diligent labourer carries a mass of pollen, kneaded by the aid of the comb at the end of the shank into a paste, which is deposited in the cells, and contributes to form the family store of provision.

What a number of compensating contrivances

does this single animal exhibit, and how wonderfully and admirably has Supreme Wisdom and Goodness contrived for her, and Almighty Power given full effect to what they planned! Nothing is superfluous in her, every hair and every angle has its use; so that well may we adore Him who created the honey-bee, and, at whose bidding, and by whose instruction, she erects those wonderful edifices that have been the admiration of every age.[1]

Instinct directs many animals, as well as traversing the surface of the earth, to seek a subterranean abode within its bosom. Amongst insects, though there are many that burrow, none is more remarkable than the *mole-cricket.*[2] The most superficial observer, when he looks at this creature, must see at once from its structure, especially that of its fore-legs, what its function is. If he compares other crickets with it, a singular change will strike him, the bulk of the posterior thighs, far exceeding that of the same joint in the other legs, will appear to be chiefly transferred to the anterior pair of legs, which, the size of the creature considered, are as powerful instruments for excavating the earth as can be found in any animal now in existence: all the joints of this leg are very much dilated, especially the haunch and the thigh, which con-

---

[1] See Bochart *Hierozoic.* ii. 515. a.        [2] *Gryllotalpa.*

tain the powerful muscles that move the apparatus for burrowing. This consists of a triangular joint, the analogue of the shank of the other legs, but assuming the form of a hand with the palm turned outwards, as in the mole, and terminating in four strong claw-like digitations; on the side next the head these fingers, in the middle, are longitudinally elevated and naked; while the sides are longitudinally excavated and hairy, which give this part some resemblance to the foot and claws of burrowing quadrupeds. The thigh is hollowed out underneath, evidently to receive the joint just described, and overhanging this cavity, at the base, is a stout triangular tooth, which probably is employed to clean the hand when necessary; on the outside opposed to the hand is the analogue of the tarsus consisting of three joints, the two first large and triangular, with the upper edge curved and the lower straight and hairy at the base, the other is of the ordinary form, and armed with two straight claws. These teeth, as well as those of the shank, have a trenchant edge on the straight side, and together are supposed to act the part of a pair of shears, and to cut any roots that may interfere with its progress. Rösel, however, thinks, the use of these teeth of the tarsus is merely to clean the burrowing hand, which it may also do. It is to be observed that the trenchant edge is opposite in the teeth of the shank and tarsus, as in a pair of scissors, which favors the idea that they are used sometimes for cutting. The position of the shank is vertical, with the teeth next the ground, so that the animal, when disposed to burrow, has nothing to do but to plunge these claws into the soil and push outwards, and then extracting her arms proceed in the same way till she has accomplished her object. The apex of the shanks, of the two posterior pairs of legs, is armed with several spines which probably assist either in making progress, or, when necessary, to retrograde.

" It might, I think, be asserted," observes Dr. Kidd, in his valuable and interesting memoir *On the anatomy of the mole-cricket,*[1] " without fear of contradiction, that throughout the whole range of animated nature, there is not a stronger instance of what may be called intentional structure, than is afforded by that part of the mole-cricket *(the anterior leg)*, which I am now to describe." And certainly, we see and own without hesitation, as even the most sceptical would scarcely refuse doing, that this arm was planned, and all its various parts, dependent upon and mutually affecting each other, by a calculating Mind, which framed and put the whole together to answer a particular purpose.

[1] *Philos. Trans.* 1825, 217.

The Class of *reptiles* affords no very striking
instances of the adaptations we are considering,
except in the case before noticed of the gecko
lizards, and the tree-frogs,[1] which, by means of
suckers, are enabled to support themselves and
walk against gravity.   Like Mammalians, reptiles are usually furnished, but not ir vari. bly,
with four legs, and a pentadactyle foot.

In an animal of this Class, celebrated from of
old, the *Chameleon*,[2] a remarkable modification
of this structure is observable.   It is stated with
respect to this animal, that it moves very slowly,
that it will sometimes remain whole days on the
same branch: and it is only with great circum-
spection, and after taking great care to get firm
hold with its prehensile tail, that it ventures to
set a few steps: it may be expected, therefore,
that its principal organs of locomotion should be
adapted to give it secure footing on the branch
it selects for its station.

Aristotle, in his account of this animal,[3] observes that " each of its feet is divided into two
parts, an arrangement resembling that of our
thumb, opposed to the rest of the hand ; and a
little short of this,[4] each of these parts is divided
into certain fingers ; in the fore-legs the internal

---

[1] *Hyla.*                [2] *Chamæleo africanus*, &c.

[3] Aristot. *Hist. Anim.* l. ii. c. 11.

[4] Gr. Επι βραχει.  Meaning, I suppose, that the toes are not
so long as the primary division of the foot.

ones being three, and the external two,[1] but in the hind the internal fingers are three, and the external two,[2] and these fingers have crooked claws." By this structure of the feet, and arrangement of the fingers or toes, the three-toed lobe is on one side of the branch at the anterior extremity of the animal, and on the other at the posterior, and by this counteraction of each other's pressure, enable it to maintain its position against any force that may be likely to disturb it. The lobes are longer than the fingers, and thus by their means it can hold very firmly, and watch the flies and insects which form its food, and are entrapped by the gluten with which its long tongue is besmeared.

The analogue of the fore-leg of quadrupeds in *birds,* as we have seen, is converted into an organ of flight, and cannot be employed as an organ of prehension; sometimes, indeed, in their combats, it is used to annoy their opponents, and is occasionally armed with a spur, but the prehensory faculty is transferred to the beak and the remaining pair of legs; with these latter the eagles and other birds of prey usually seize the animals that they devour; with these also fructivorous birds, as the parrots, paroquets, &c. hold the fruit while they eat it, and the Gallinaceous

[1] PLATE XIV. FIG. 2.  [2] *Ibid.* FIG. 3.

Order scratch the earth to find food for themselves and chicks; the foot of birds is most commonly tetradactyle, with one toe or thumb at the heel and the other three in front; in one Order,[1] the birds forming which have occasion to fix themselves firmly on their perch, the thumb and the external toe both point backwards, so as to form a cross with the others and the rest of the leg. In the cmu the foot consists of three toes, and in the ostrich of only two, there being no thumb in either. Many of the aquatic birds have the toes connected by membrane, and so forming oars for swimming; and in some each toe has a margin of membrane, which is usually notched, these last are called lobed feet.

But the absence of the fore-leg in birds is admirably compensated by the *beak*; with this they generally *collect*, as well as devour their food. Some indeed employ their *tongue* in this service. Of this description is the woodpecker[2] and the humming Bird;[3] the former using it to catch insects[4] and the latter to imbibe the nectar of flowers, for which purpose these little gems amongst the birds have a long slender tongue, somewhat resembling that of a butterfly, and moved by an apparatus, in some degree, like

[1] *Scansores.*          [2] *Picus.*
[3] *Trochilus.*          [4] See Dr. Roget, *B. T.* ii. 132.

that of the woodpecker.[1] The beak of birds is uniformly constructed with respect to their food, and varies ad infinitum. Perhaps in none is it more remarkable than in those of Cuvier's two last Orders, the waders and web-footed birds. These, especially the last, can use their legs only for locomotion, either on shore or in the water, and therefore their beaks have the whole function, not only of taking, but of *hunting* for food devolved upon them, and accordingly are fitted for it by their structure.[2] Generally speaking, they may be stated to be of two kinds. Beaks for catching *worms*, and beaks for catching *fishes*; of the first description are those of the woodcock,[3] snipes,[4] and numerous other waders; and of the last, amongst the most remarkable, are those of the spoonbill[5] and pelican.[6] The former—which the French, perhaps with more propriety, call the spatula-bill,[7] as its beak resembles a spatula rather than a spoon—dabble with their bill in the mud, for which it is well calculated, and thus capture small fishes, shell-fish, reptiles, and other aquatic and amphibious animals, which the tubercles within it are also calculated to retain and crush. But the

---

[1] See Vieillot. *N. D. D'Hist. Nat.* vii. 342. *t.* B. 38.

[2] Roget, *B. T.* ii. 391.      [3] *Scolopax rusticola.*

[4] *Sc. gallinago,* and *gallinula.*      [5] *Platalea leucorodia.*

[6] *Pelecanus Onocrotalus.*      [7] *Spatule.*

latter, the pelican, has the most remarkable organ for taking its food, and is a bird known and celebrated from the earliest ages. The lower mandible is fitted with a kind of sac, formed of the dilated skin of the throat, which Vieillot says can be so expanded as to contain between two and three gallons of water.[1] When fishing these birds, sometimes, rise to a prodigious height, at others they skim the surface of the water, or hover, at a moderate elevation, that they may more readily precipitate themselves upon their prey. The sudden fall of so powerful an animal, the whirling round, the boiling which the great extent of its wings occasions in the water, so astounds and stuns the fishes that few escape. Then rising again and again descending, it continues this manœuvre till it has filled its pouch. When this is accomplished it retires to some rocky eminence where it devours what it has caught, which sometimes, Vieillot says, will amount to as many fishes as would satisfy six men.[2] It presses its pouch against its breast when it feeds its young, in order to disgorge the fishes, whence probably arose the fable of its feeding them with its own blood.

But the beak is not only used by birds in collecting their food, some also it assists in

---

[1] *N. D. D'Hist. Nat.* xxv. 139.    [2] *Ubi supr.* 138.

*climbing;* parrots are remarkable for this, and also employ their *tail* for the same purpose.

Truly, when we examine and compare all these organs of prehension as well as manducation, and the infinite modifications of them, to suit the peculiar kind of food and circumstances of every tribe, we cannot help exclaiming—God is here, we behold the evident footsteps of infinite wisdom, power, and goodness. Well might our Saviour say, *Behold the fowls of the air; for they sow not, neither do they reap, nor gather into barns; yet your Heavenly Father feedeth them.*[1]

The legs of *Mammalians*, with respect to their extremity, may be considered as divided into those that have powers of prehension, more or less, and those that have only powers of locomotion. I shall begin with the latter.

1. These consist of Baron Cuvier's *seventh and eighth* Orders of the Class above mentioned; namely, the *Pachyderms*, or thick skinned beasts, and the *Ruminants*, or those that chew the cud.

The great man, just named, considers the horse and ass, constituting the equine genus,[2] as forming a Family of the first of these Orders, to which he has given the ancient appellation of

---

[1] *Matth.* vi. 26.  [2] *Equus.*

*Soliped*,[1] or whole-hoofed.  He originally regarded the Solipeds as forming a separate Order,
and, indeed, comparing them with the other
Pachyderms, as the elephant, rhinoceros, hippopotamus, hog, &c., the horse genus seems
scarcely to belong to the same Order.  Illiger,
who altered the name, but without sufficient
reason, to *Solidungula*, considers them as distinct.

Though the speed of the deer, except in a
single instance, on account of their usually slight
form and slender limbs, has not been applied by
man to his purposes, and to add to the velocity
of his progress, yet in the soliped race, especially in that noble quadruped the *horse*, we
have an animal endowed with equal speed and
greater strength, and by their undivided hoof,
where speed as well as strength is required, calculated, with much more advantage and less injury, to traverse—both as beasts of burthen and
draft, and as adapted peculiarly for the conveyance of man himself—not only soft and
verdant prairies, but hard and rocky roads.
Hence this animal has been employed by man
from a very early period of society.  We do not
indeed know whether the mighty hunter, Nim-
rod, went to the chase of man and beast on
horseback, though it is not improbable ; but both

---

[1] Gr. Μονυξ.　Aristot.

the horse and the ass were common in Egypt in Joseph's time,[1] the latter was used by Abraham to ride upon,[2] and asses are enumerated amongst his possessions when he went up from Egypt fifty years before.[3]

The sole organs of prehension of this tribe are their mouth and upper lip. Every one knows how adroit the horse and ass often become in the use of these organs, not only in gathering their food, but in opening gates that confine them to their pastures.

In the genuine Pachyderms the foot begins to show marks of division. In the rhinoceros there are three toes, in the hippopotamus four, and in the Proboscidians of Cuvier, including the elephant and *Mastodon*, or fossil elephant, there are five toes, three of the nails of which only appear externally, and four on the hind-foot of the Asiatic species.[4]

The *Swine* family divide the hoof like the Ruminants; it consists of two intermediate toes, large, and armed with nails or hoofs, and two lateral ones much shorter and not touching the ground; in this respect also resembling many Ruminants. In hilly and mountainous districts these upper toes are probably useful in locomotion.

The prehensory organ of the animals here

[1] *Genes.* xlvii. 17.　　[2] xxii. 3.　　[3] xii. 16.　　[4] *E. indicus.*

enumerated is usually the *snout*, with this the *hog*[1] turns up the ground in search of roots or grubs, often doing great injury to pastures. The male is armed with a defensive and offensive weapon in his tusks.

That hideous animal of this tribe, the *Æthiopian boar*,[2] is armed with four tusks, two proceeding from the upper jaw, which turn upwards like a horn, sometimes nine inches long and five inches in circumference at the base; the other pair issuing from the lower jaw, projecting not more than three inches from the mouth, flat on the inside, and corresponding with another plain surface in the upper tusks. The Boshies men, Sparrman relates, say of this animal, " We had rather attack a lion in the plain than an African wild boar; for this, though much smaller, comes rushing on a man as swift as an arrow, and throwing him down snaps his legs in two, and rips up his belly before he can get to strike at it, and kill it with his javelin."[3] They inhabit subterranean recesses; and turn up the earth very dexterously, probably by the aid of their tusks, in search of roots, which form their food.

The *Babiroussa*[4] or *Babee rooso*, a name which signifies *Hog-deer*, given to this animal probably on account of its longer legs and slender form, is distinguished by a pair of long tusks from the

---

[1] *Sus scrofa.*    [2] *Phascochœrus Africanus.*
[3] Voyage, ii. 23.    [4] *Sus babyrussa.*

upper jaw, which rising above the head, then turning down, form a semicircle, and have the appearance of horns, for which they have been mistaken. They are only found in the male, which is stated to use them as hooks to suspend himself to the branches of trees, thus resting his head, so as to sleep upright. As the animal feeds upon the leaves of the Banana and other trees, it is not improbable that these tusks may be used to pull down the branches.

The *Rhinoceros* is said to use its horn for digging up the roots of plants, which compose the principal portion of its food. I am speaking of the two-horned rhinoceros of Sparrman. The Hottentots and the colonists assert that this animal uses only its *second* or shortest horn for digging up roots, which appeared to him worn by friction, marks of which the anterior one never exhibited. When engaged in that employment it was stated to turn that horn on one side[1] out of the way.

But one of the most wonderful compensating contrivances and structures of Divine Wisdom, Power, and Goodness, and which has excited the admiration of every age, is the *proboscis* of the *elephant*. The weight of the enormous head of this animal is such as to preclude its being employed, if it terminated in a common mouth,

[1] Sparrman. *Voyage.* ii. 98.

either to break the boughs of trees, or to crop the grass, for it could not easily be either elevated or depressed for these purposes; in its proboscis, however, it is supplied with an instrument that amply compensates this deficiency. Almost every one is aware that this beautiful organ, beautiful I mean for its structure,[1] answers a variety of purposes; that it is given by its Creator to this mighty animal to be to it an instrument almost of sight, of most delicate touch, of scent and breathing, of prehension as adroit as that of a hand; added to this, that by the extraordinary flexibility with which he has endowed it, it can not only be inflected inwards to carry things to its mouth, but be bent upwards, downwards, or laterally, to lay hold of things above, below, or on each side of it, and that by the assistance of a single finger at its extremity, it can take hold of any thing as readily as we do by the assistance of four fingers and a thumb. As the brain of these gigantic animals, compared with their bulk, is very small, it is thought, by modern zoologists, that their intellect has been exaggerated, and that it does not surpass that of dogs, and many other carnivorous animals. Others have imagined that their sagacity is wholly the result of their being provided with so wonderful an organ; but this organ would be of very little use without

---

[1] Roget, *B. T.* i. 520.

the *nervous* apparatus by which it is moved according to the will of the animal.

Amongst the *Ruminants*,—which appear to connect with the Pachyderms in two points, by the swine tribe and Solipeds, the latter possessing several characters in common with the *Gnu*,[1] which seems between them and the bovine genus;[2] and the former approaching them by their common character of dividing the hoof,—there is another animal, which may be considered as the horse of the desert, exhibiting in some degree a union of characters not found in the remainder of the Order; it chews the cud, but does not actually divide the hoof. I am speaking of the *Camel*, but though not actually, the hoof is superficially divided. Considering the deserts of loose and deep sand that it often has to traverse, a completely divided hoof would have sunk too deep in the sand; while one entire below would present a broader surface not so liable to this inconvenience. Boys, when they want to walk upon the muddy shores of an estuary at low water, fasten broad boards to their feet, which prevent them from sinking in the mud; I conceive that the *whole* sole of the camel's foot answers a similar purpose: its superficial division probably gives a degree of pliancy to it, enabling it to move with more ease

---

[1] *Catoblepas Gnu.*			[2] *Bos.*

over the sands; upon which these animals often trot with great rapidity, travelling sometimes twelve miles within the hour; its common amble, which is exceedingly easy, is nearly six; this pace, if properly fed every evening, or in cases of emergence, only once in two days, the camel will continue uninterruptedly for five or six days: with these qualities, so suitable to barren and sandy deserts, what a valuable gift of Providence was this, especially to the descendants of Ishmael; who, according to the prophecy, have maintained undisturbed possession of their deserts and their necessary accompaniment, the camel, from the time of their progenitor to the present day, a period of between three or four thousand years. They have been wild men, always assailing and assailed, and yet maintaining their ground. But the time will assuredly come, when *The flocks of Kedar, and the rams of Nebaioth,*[1] shall forsake their deeds of spoliation and robbery and be gathered to the church.

Though the Ruminants, in general, by the structure and division of their hoof, are calculated for sure footing, so as to enable them best to exercise their several functions; as the camel, the ox, and the rein-deer at the bidding of their master man; and others, as the chamois and the goat, for the ascent of mountains and pre-

---

[1] *Isai.* lx. 7.

cipices, seemingly inaccessible, where they can
laugh at their pursuer ; and others again, as the
deer and antilope tribes for speed that almost
mocks pursuit ; yet with respect to *prehension*
these organs are of no use to them. Their mouth
and lips, and tongue, are the only means by
which they can help themselves to their food ;
they have no tusks like the Pachyderms in
general, nor nasal horns like the rhinoceros, to
cut or dig with ; but as their food is most commonly the herbage that covers the earth, these
are fully sufficient to enable them to supply
themselves with *Food convenient for them*. The
camel and dromedary differ from the other
Ruminants, not only in their long neck, which
probably is useful to them in gathering their
food, but also in having a cleft lip, which doubtless, adds to the prehensory powers of that
organ. The lofty neck is still more striking in
the Camelopard, the long tongue of which is
also used by them as a hand to pull down the
branches of the mimosa, from which they derive
their subsistence.

2. I shall now consider those Mammalians,
whose legs are more or less prehensory, next
above the Pachyderms and Ruminants. Cu
vier's *sixth* Order consists of a tribe of animals
which he denominates *Edentate*,[1] because they

---

[1] *Edentés.*

have no fore-teeth. The *Monotremes* form the last Family of the Order, and precede the Pachyderms. In many points they seem connected with the birds; one genus[1] having a mouth resembling the bill of a duck, and being almost web-footed; it has also been stated to be oviparous;[2] the male, as I before observed,[3] is armed with a sting, like a serpent. The other genus, *Echidna*, approaches nearer the *pangolins*,[4] and *anteaters*,[5] having, like them, an extensile viscid tongue, by means of which they entrap and devour the ants. The other animals of the Order are remarkable for their great nails, almost approaching to hoofs; in the Family which precedes the Monotremes[6] they are often used for burrowing.

Next above the *Echidna* is a singular animal, wearing the outward aspect and scales of a Saurian, the pangolin, which rolls itself up like an armadillo, and is the ant-eater of the old world. It is singular that a real lizard, the chameleon, should have the same instinct of catching its insect prey by means of a long tongue besmeared with slime. In the new world the pangolin is replaced by the ant-eaters, which have the same habits, and the same mode of

---

[1] *Ornithorhynchus.*      [2] Cuv. *Règne Anim.* i. 234, note 2.
[3] See above p. 82.       [4] *Manis.*
[5] *Myrmecophaga.*         [6] *Edentés ordinaires.*   Cuv.

procuring their food. With the long nails of their fore-feet they penetrate the nests of the white ants and common ants, and inserting their long tongue, besmeared with a viscid saliva, into these nests, retract it covered with game ; and this with such velocity, that the eye can scarcely follow them. Their nails, which require to be kept sharp, for the operation just mentioned, when not employed, are folded inwards, so as to prevent their being blunted. In one species[1] in the fore-foot there are only two nails.

Amongst the animals that are clothed in armour, in this Order, the most remarkable is the *Chlamyphorus*,[2] whose feet are armed with five long and sharp nails, especially the anterior ones, which must enable it to excavate its subterranean abode very rapidly. From the formation of its foot and these nails it does not appear to dig with them laterally, but in a line with the body ; its singular clubbed tail therefore would be a very useful organ, if, as Mr. Yarrell supposes, it is used in removing backwards the loose earth accumulated under its belly by the action of the fore-legs.[3] This animal, which is a native of Chili, is reputed to carry its young beneath the scaly armour attached principally to the spine, which covers it loosely like a cloak.

[1] *M. didactylas.*　　[2] PLATE XVI.　　[3] *Zool. Journ.* iii. 551.

The last family, as we ascend, in the present Order, is very well distinguished by the name of *Tardigrades*, from the excessive slowness of their motions. Their nails are enormously long, compressed, and crooked, and exactly calculated for laying strong hold, so as to enable them to maintain their station on the trees, whose leaves and buds form their food. Their English appellation, the *Sloth*,[1] indicates their character; when they have satisfied their appetite, like most of the other Edentates, they can roll themselves up and take a long and reckless sleep. But I need not enlarge further upon this tribe, since Dr. Buckland has excellently—*Justified the ways of God to man,*—and, in the present instance, demonstrated, by most convincing arguments, that these animals, instead of being an abortion, imperfect, misshapen, and monstrous, are exactly, and in every respect, adapted for the station which God has assigned to them, and for the work which he has given them in charge.[2]

Next above the Edentate Mammalians is an Order, the *fifth* of Cuvier, consisting of a greater number of Genera and Subgenera than any other in the Class, which, instead of having no front teeth or incisives, have very conspicuous ones, rendered more so by being separated by a

---

[1] *Bradypus.*　　　[2] *Linn. Trans.* xvii. 17.

void space from the grinders. From these teeth, which are neither calculated to seize or lacerate their food, but merely to nibble and gnaw it, they have received their name of *Nibblers* or *Gnawers*.[1]

The great majority of this Order are gregarious, and live in burrows, or common habitations, which they excavate or fabricate themselves. Like the Hymenopterous Class of insects, many are noted for the sagacity and skill which they manifest in their united labours for the good of the community, and also for the organs by which they are enabled to answer the bidding of instinct.

One of the most remarkable of these is the *Beaver*;[2] this animal has five toes on all its feet, which in the hind pair are connected by membrane; those of the fore-leg, which it uses as a hand to convey its food to its mouth, are very distinct. They carry also with these hands the mud and stones which they mix with the wooden part of their buildings. But their incisor teeth are their principal instruments, with these, as Dr. Richardson states, they cut down trees as big or bigger than a man's thigh; when they undertake this operation they gnaw it all round, cutting it sagaciously on one side higher than on the other, by which it is caused to fall in the

---

[1] *Rodentia.*  [2] *Castor Fiber.*

direction they wish; they use these powerful organs not only to fell the trees they select, but also to drag them to the place where they want them. It is said, that a beaver, when at its full strength, can at one stroke bite through the leg of a dog.

It has been affirmed that beavers employ their tail both as a trowel to plaster their houses, and as a sledge to carry the trees that they fell; but both these assertions seem to be built upon conjecture rather than observation, and are not credited by those who have had the best opportunities of observing their manners, as Hearne, Cartwright, and Dr. Richardson. The fabrics they are taught by their Creator to erect, and impelled by the instinct he has implanted in them, are sufficiently wonderful without having recourse to fiction to exaggerate it. Their tails, probably, are useful to them in the water as natatory organs.

There is a very singular animal discovered by M. Sonnerat, in Madagascar, called the *Aye-Aye*,[1] which seems, in some degree, to approach the Quadrumanes. The fore-feet have five excessively long fingers, and what is singular, the middle one is much slenderer than the rest. In the hind feet there is a thumb opposed to the other fingers, by which structure it is enabled to

---

[1] *Cheiromys.*

take firmer hold of the branches of trees. It is said to use the slender finger of its hand for the same purpose that the wood-pecker uses its barbed tongue, to extract the grubs from the trees.

The squirrels, which form the first genus in this interesting Order, are known to use their fore-legs for prehension, which indeed is the case with the majority of animals included in it. They are also, at least a large proportion, remarkable for sitting, when at rest, upon their haunches, and also for their ready use of their fore-legs.

Having before noticed the most remarkable animal in Cuvier's *fourth* Order, the *Marsupians*, which suckle their young in a pouch, I shall only mention one other animal belonging to it, the *Koala*,[1] a New Holland quadruped, in some respects resembling the bear; like the chameleon, it has the five toes or fingers of the fore-foot divided into two groups, the thumb and fore-finger forming one, and the three remaining fingers the other; the object of this structure is evidently to enable it to take firm hold of the branches of the trees on which it passes part of its life; this is of the more importance to it, as it carries its young upon its back. It some-

[1] *Lipurus.*

times, probably in the night, retires to burrows which it excavates at the foot of the trees.

We have now arrived at the foot of Baron Cuvier's *third* Order, containing the *predaceous* Mammalians, which, though a very comprehensive group, will not detain us long, as the first and last family, the *Bats* and *Seals*, have been noticed in another place.[1] The rest of the Order consists of the insectivorous and carnivorous Mammalians; the latter is further subdivided into two tribes, which are denominated the *Plantigrades* and the *Digitigrades*.

Those last mentioned usually walk more upon their toes, and consist of the feline, canine, and several other tribes, all swift in their locomotions, and making use of their paws or fore-foot, either for scratching and burrowing, or to seize their prey, and they have all, I believe, five toes.

The *Plantigrades* are so called because they walk, like man, upon the whole foot, and consist of the bear,[2] the glutton,[3] and similar animals. This structure enables the former to rear itself on its hind feet, and walk erect; and their forefoot will grasp a staff like a hand; it is armed with long claws, with which they scratch up roots which form part of their subsistence, exca-

---

[1] See above, p. 137, 156.     [2] *Ursus.*     [3] *Gulo.*

vate burrows, climb the trees, and seize their prey.

These armed paws are fearful weapons, both in the lion and the bear, to which few would like to be exposed ; but an heroic youth, beloved of God and man, regarded them not when, as a faithful shepherd, he rescued a lamb of his father's flock from their grasp and voracity.

The two most remarkable animals in the *insectivorous* tribe of predaceous Mammalians are the mole,[1] and the harmless, though persecuted hedgehog,[2] but they are both too well known, the former for its piquants, and the latter for its hand turned outwards and moved by an enormous apparatus of muscles, to enable it to excavate its subterranean habitation.

We are now arrived, in our progress upwards, at Cuvier's *second* Order of Mammalians, which he names *Quadrumane*, or four-handed, and which consists of apes,[3] baboons,[4] and monkeys,[5] whose hind as well as fore-foot is usually furnished with a thumb opposite to the fingers, so that they can use all their feet for prehension : the object of Providence by this structure is to enable these animals to move about amongst the branches of the trees, which are their usual habitations, and to fix themselves

---

[1] *Talpa.*      [2] *Erinaceus.*      [3] *Simia,* &c.
[4] *Cynocephalus,* &c.      [5] *Lemur,* &c.

securely upon them, so that they can use their hands to gather fruit or any other purpose. Thus also they can perambulate the trees with as much ease and safety as we do our houses; and run up and down the branches with as much celerity as we do our staircases : but they cannot make equal progress on the earth, or a plane surface, whether they go on four feet or two.

Even man himself, though he ordinarily cannot use his toes for prehension, yet is sometimes placed in such circumstances, as to acquire the power of doing so. I remember, when a boy, going to see a girl who was born without arms, and was exhibited by her parents to the public. She could use her toes as fingers; could hold scissors, cut out watch-papers, sew, and even write. An account was given in the St. James's Chronicle, not long ago, of a youth similarly circumstanced, who being cruelly turned out by his father, but patronized by his sister, learned to draw with his toes. In India they are used as fingers, and are sometimes called foot-fingers. The Hindoo tailor twists his thread with them, and the cook holds his knife while he cuts fish, vegetables, &c., the joiner, weaver, and other mechanics all use them for a variety of purposes ; and I am told by a friend, who has often been in India, that they can even pick up pins with them.

We are now arrived at man himself, who, as we see, takes his particular denomination from the hand.   He is the only *Bimane.*

The physiology and anatomy of the *Human Hand*, that wonderful organ, have been explained and reasoned with great ability in a separate treatise, by the eminent comparative anatomist to whom that subject was assigned; I shall not, therefore, here say any thing on its structure and its uses: but as it has not been treated of as a *moral* organ; as being in intimate connection with the heart and affections; as their principal index and premonstrator; and as the mighty instrument by which a great part of the physical good and evil which befalls our race is wrought, I may be permitted to make a few observations upon it as far as these are concerned.

God made the body in general a fit machine, not only to execute the purposes of its immaterial inhabitant the soul; but, in some sort, he made it a mirror to reflect all its bearings and character; to indicate every motion of the fluctuating sea within, whether its surges lift themselves on high elevated by the gusts of passion; or all is calm, and tranquil, and subdued.   None of the bodily organs, by its structure and station in the body, is so evidently formed in all respects for these functions as the HAND.   The eye indeed is, perhaps, the most faithful mirror of the soul's emotion; yet though it may best portray and render visible the internal feeling, it can in no degree execute its biddings; but the hand is the great agent and minister of the soul, which not only reveals her inmost affection and feeling, and, in conjunction with the tongue—and these two in connection are either the most beneficent or maleficent of all our organs—declares her will and purpose; but is also employed by her to execute them. Thus HEART and HAND, the principle and the practice, have been united, in common parlance, from ancient ages. The earliest dawn of reason in the innocent infant is shown by the signs it makes with its little *hands;* by them it prefers its petitions for any thing it desires, and, in imitation of this, God's children are instructed *to lift up holy hands* in prayer.[1] Love, friendship, charity, and all the kindly affections of our nature, use the hands as their symbol and organ; the fond embrace, the hearty shake, the liberal gift, are all ministered by them. Joy, gladness, applause, welcome, valediction, all use these organs to represent them. Penitence smites her breast with them; resignation clasps them; devotion and the love of God stretches them out towards heaven.

But the hands are not employed to express

---

[1] 1 *Tim.* ii. 8.

only the kindly affections of the soul. Those of
a contrary and less amiable character use them
as their index. Anger threatens, and more
violent and hateful passions destroy by them.
They are indeed the instruments by which a
great portion of the evil, and mischief, and
violence, and misery, that our corrupt nature
has introduced into the world, are perpetrated.

The hand also, on some occasions, becomes
the spokesman instead of the tongue. The fore-
finger is denominated the *index*, because we use
it to indicate to another any object to which we
wish to direct his attention. By it the deaf and
dumb person is enabled to hold converse with
others so as not to be totally cut off from the
enjoyment of society; and by it we can likewise mutually communicate our thoughts when
separated by space however wide, even with our
Antipodes.

The Deity himself, also, condescends to convey spiritual benefits to his people by means
of the *hands* of authorized persons, as in Confirmation and Ordination; and the Blessed
Friend, and Patron, and Advocate and Deli-
verer of our race, when he was upon earth,
appears to have wrought most of his miracles of
healing by laying on his hands;[1] in benediction
also, when children were brought unto him he

---

[1] *Mark*, viii. 23—25.

laid his hands on them ; and at his ascension he lifted up his hands to bless his disciples.[1]

To enumerate all the modes by which the internal affection of the soul is indicated by the hand would be an endless task. I shall therefore only further observe, that the greater part of the instances I have adduced are natural, and not conventional or casual modes of expressing feeling, as is evident from their being employed, with little variation, in all ages, nations, and states of society.

How grateful then ought we to be to our Creator for enriching us with these admirable organs, which more than any outward one that we possess, are the immediate instruments that enable us to master the whole globe that we inhabit, not merely the visible and tangible matter that we tread upon, and its furniture and population, but even often to take hold as it were of the invisible substances that float around it, and to bottle up the lightning and the wind, as well as the waters. Thus by their means do we add daily increments to our knowledge and science, and consequently power ; to our skill in arts and every allied manufacture and manipulation ; to our comforts, pleasures, and every thing desirable in life.

If now—having arrived at the most perfect

---

[1] *Mark*, x. 16. *Luke*, xxiv. 50.

instrument, as to its uses, and the most important to the happiness and welfare of the Human race, whether it be considered as an instrument of good or evil—we turn back and review this long train of organs for every kind of motion, and every kind of operation, and consider moreover the animal to which each belongs with respect to its place and station, connection, powers of multiplication, relative magnitude, form, composition, structure, functions, and at the same time take into further consideration the theatre upon which each is destined to appear, the medium in which it is to move and breathe, and the beings, whether vegetable or animal, with which it is to come in contact, and upon which it is to act.

When, I say, we take this review, what an infinite diversity in every respect bewilders our thought, and we are unable to form any distinct idea of the general effect and harmony that we know to be produced, nor how all these instruments, dove-tail, as it were, so as to form the whole into one great fabric or sphere of agents, all contributing to fulfil the purposes of the Great Being who fabricated it, and promoting the general health and welfare of the whole system. But this we *can* understand that the Fabricator of this sphere must have taken a *simultaneous* survey of all the circumstances here mentioned ; must have calculated the momentum of each

individual, have weighed and measured it, so that
it should not exceed a certain standard ; must
have seen at once all that it wanted to fit it for
its station ; must, before he made it, have formed
a correct estimate of all the requisite materials,
whether gaseous, aquiform, or solid, so as to put
together the whole harmonious compages without failing in a single atom ; and give full
accomplishment to his will.

He who could effect all this could only be
one whose *Understanding is infinite,* and whose
*Power and Goodness* are equally without bounds.

CHAPTER XVIII.

*On Instinct.*

THERE is no department of Zoological Science
that furnishes stronger proofs of the being and
attributes of the Deity, than that which relates
to the *Instincts* of animals, and the more so,
because where reason and intellect are most
powerful and sufficient as guides, as in man,
and most of the higher grades of animals, there
usually instinct is weakest and least wonderful,
while, as we descend in the scale, we come to
tribes that exhibit, in an almost miraculous

manner, the workings of a Divine Power, and perform operations that the intellect and skill of man would in vain attempt to rival or to imitate. Yet there is no question, concerning which the Natural Historian and Physiologist seems more at a loss than when he is asked—what is Instinct? So much has been ably written upon the subject, so many hypotheses have been broached, that it seems wonderful so thick a cloud should still rest upon it. It must not be expected, where so many eminent men have more or less failed, that one of less powers should be enabled to throw much new light upon this palpable obscure, or dissipate all the darkness that envelopes the *secondary* or intermediate *cause* of Instinct. Could even the bee or the ant tell us what it is that goads them to their several labours, and instructs them how to perform them, perhaps we might still have much to learn before we should have any right to cry with the Syracusan Mathematician, Ευρηκα, I have unveiled the mystery. Still, however unequal to the task, I cannot duly discharge the duty incumbent upon me, who may be said to be *officially* engaged to prove the great truths of Natural Religion from the *Instincts* of the animal creation, to leave the subject of Instinct, considered in the abstract, exactly as I found it; a field, in which whoever perambulates, may wander " in endless mazes lost."

I will, therefore, do my best to make the way, in a small degree, more level, and less intricate, than it has hitherto been.

But, before I proceed, lest the reader should feel disposed to accuse me of contradicting the opinions on this subject stated in the *Introduction to Entomology*, I beg to direct his attention to the following paragraph in the advertisement to the third volume of that work. " It will not be amiss here to state, in order to obviate any charge of inconsistency in the possible event of Mr. Kirby's adverting in any other work to this subject, that, though on every material point, the authors have agreed in opinion, their views of the *theory of Instinct* do not precisely accord. That given in the second and fourth volumes is from the pen of Mr. Spence."

It is not without considerable reluctance that the author of this essay takes the field, in some degree, against his worthy friend and learned coadjutor, but as he is thus left at liberty to do it, and the nature of his subject requires it, he will state those views, which seem to himself most consistent with nature and truth, and most accordant with the general plan of creation. It is doubtful whether the ancients had any distinct idea of that impulse upon animals, urging them necessarily to certain actions, which modern writers have denominated *instinct*.

Aristotle, indeed, in a passage of his physics quoted by Bochart,[1] alludes to certain writers who doubted whether spiders, ants, and similar animals were directed in their works by intellect, or by any other faculty. The Stagyrite himself resolves the causes of motion into intellect and appetite,[2] but I have not been able to discover that he has recorded any opinion as to what cause the, now called, instincts of animals, whether to appetite or intellect, are to be attributed : he says much on the subject of the hive bee, but it is merely a history of its proceedings, unaccompanied by a single syllable from which we might conjecture that he attributed any part of these proceedings, wonderful as he must have thought them, to any faculty distinct from intellect, and what seems more extraordinary, without any expression of admiration at the expertness, and art, and skill, so evident in all that this little creature almost miraculously accomplishes. On another occasion, indeed, he observes, that " Some of the animals that have no blood, have a more intelligent soul than some of those that have blood, as the bee and the ant genus."[3] A much later Greek writer has asked the question, " *Who taught the bee, that wise workman, to act*

[1] *Hierozoic.* ii. 599, b.
[2] *De anima,* l. iii. c. 11.
[3] *De Part. Animal.* l. ii. c. 4.

*the geometer, and to erect her three-storied houses of hexagonal structures?*[1]   And this is the question I shall now endeavour to answer.

When we consider the infinite variety of instincts, their nice and striking adaptation to the circumstances, wants, and station of the several animals that are endowed with them, of which numerous instances will be given hereafter, we see such evident marks of design, and such varied attention to so many particulars, such a conformity between the organs and instruments of each animal, and the work it has to do, that we cannot hesitate a moment to ascribe it to some power who planned the machine with a view to accomplish a certain purpose, and when we further consider that all the different animals combine to fulfil *one* great end, and to effect a vast purpose, all the details of which the human intellect cannot embrace, we are led further to acknowledge that the whole was planned and executed by a Being whose essence is unfathomable, and whose power is irresistible.

I must here previously observe, that in considering this mysterious subject, we must avoid, as much as possible, building our theories upon facts which, if properly interpreted, are extra-

---

[1] Τις την μελιτταν, την σοφην την εργατιν
Γεωμετρειν επεισε, και τριωροφες
Οικες εγειρειν ἐξαγονων κτισματων.
          Pisidius, *De Mundi Opificio*, quoted by Bochart.

neous to the subject, and wear such an aspect of
the marvellous, as to appear out of the regular
course of nature, and the ordinary proceedings
to which its instinct urges any animal. The
cases here alluded to, if true, to the full extent
of the statements concerning them, would rather
indicate a particular interposition of Divine
Providence, either to prevent some calamity, or
to produce some blessing or benefit to the individuals concerned. Thus the account of Sir H.
Lee's dog, mentioned by Mr. French,[1] which
saved its master's life, by taking and maintaining
its station, which it had never before done, under
his bed; and that given by Dr. Beattie, of a dog,
who, when his master was in a situation of the
most imminent peril, after fruitlessly attempting
to save him, ran to a neighbouring village, and
by significant gestures at last prevailed upon a
man to follow him, and saved his master's life.
These and many more such cases, can scarcely
be regarded as belonging to the ordinary instinct
of the species, for if it did, more murderers would
be disappointed of their intended victim by the
agency of his or her dog. I knew myself an
instance, in which a most valuable life was saved
by a dog, which, being condemned to the halter
by a former master, and escaping from those

[1] *Zool. Journ.* i. 7.

appointed to dispatch him, at last established himself, after repeated expulsion, in my friend's family, and afterwards, there is every reason to believe, by the sacrifice of his own life, prevented his master from being drowned.[1] These cases are remarkable, but they do not appear to belong to instinct, but rather to the doctrine of a particular Providence.

Some cases upon record, with respect to dogs and other animals, belong to intellect and memory rather than instinct. M. Dureau de la Motte, in a memoir on the influence of domesticity in animals, mentions a dog, which being shut out, would use the knocker of the door;[2] and I had myself a cat, which indicated its wish to come in or go out, by endeavouring with its fore paws to move the handle of the door-latch of the apartment; and used every morning to call me by making the same indication at the door of my bed-room: other cats have attempted to ring the bell. But the most remarkable instance, is one related, by the writer just named, of a very intelligent dog, which was employed to carry letters between two gentlemen, and never failed punctually to execute his commission—first delivering the letter, which was fastened to his collar, and then going to the kitchen to be fed. After this, he went to the

---

[1] *Annal. des Sc. Naturel.* xxi.     [2] *Ibid.* 52.

parlour window, and barked, to tell the gentleman he was ready to carry back the answer.[1]

The remarkable case of the ass Valiante,[2] and
of other animals that find their way to their old
quarters from a great distance, may be attributed,
I think, rather to natural sagacity and memory,
than to any instinctive impulse. The animal
just alluded to might have sagacity enough to
keep near the sea, or a concurrence of accidental
circumstances might befriend her.

Divine Providence has at its disposal the
whole animal creation, and can employ all their
instincts and their faculties to bring about its own
purposes, both with respect to individuals and
mankind in general. Man, who may be called,
under God, the king of the visible creation,
makes a similar use of the creatures that are
placed at his disposal; of some, as the horse and
the ox, he employs the physical powers; of
others, as the bee and the dog, he avails himself
of the instinct. Some he instructs how they are
to do his work; others, he takes as he finds them.
So the Deity, it may be presumed, with a secret
hand, guides some to fulfill his will, instructing
them, as it were, because their unaided instinct
would not alone avail, in the decree they are to
execute, while others, merely by following the
bent of their nature, do the same. In many

[1] *Annal. des Sc. Naturel.* 66.
[2] *Introd. to Ent.* ii. 496. Note a.

cases, also, he may be supposed merely to direct them to the field in which he means they should labour, and then leave them to their instincts to accomplish his purposes.  In the case of the dog who saved his master from intended assassination, a supernatural impulse might carry him to his chamber and cause him to maintain his station there, and when the hour of danger arrived, his natural instinct would suffice for the defence and liberation of his master from the threatened danger.

When we consider the work that animals have to do in this globe of ours, each, in a particular department, and to a certain extent, it seems absolutely necessary that, on many occasions, the interference of a Supreme Power should take place, to say to each, " *Hitherto shalt thou come and no further*," and only an Omnipresent Being, infinite in power, wisdom, and goodness, could check the further progress of any body of his workmen when he foresaw it would be noxious, exceed his intentions, and derange his plans.

> " *Nec Deus intersit, nisi dignus vindice nodus*
>     *Inciderit*,"

was the dictum of a poet, who had as much judgment, and good sense, as he had genius; and it is only where ordinary means are evidently insufficient to account for any fact, that we are at liberty to ascribe it to the extraordinary interposition of the Deity ; or to any *intermediate* supernatural agency employed by him to produce it : and no class of facts so loudly proclaim their Great Author as those which are the result of the nice balancing of conflicting energies and operations observable in the different departments of the animal kingdom.

We may observe, however, that when our Saviour says to his disciples concerning sparrows—*One of them shall not fall to the ground without your Father. But the very hairs of your head are all numbered ;*[1]—the observation implies that nothing escapes the notice, or is too mean, or insignificant, to be below the attention and care of Him who is all eye, all ear, all intellect ; who directeth all things to answer his purposes, *according to the good pleasure of his will,*[2] which is the universal good of his creatures.

Having premised these general observations, I shall now proceed to inquire into the proximate cause of instinct; admitting, as proved, that every kind of instinct has its origin in the will of the Deity, and that the animal exhibiting it, was expressly organized by Him for it at its creation.

The proximate cause of instinct must be either metaphysical or physical, or a compound of both characters.

1. If *metaphysical*, it must either be the *im-*

---

[1] Matth. x. 29, 30.        [2] *Ephes.* i. 5.

*mediate* action of the Deity, or the action of some *intermediate* intelligence employed by him, or the *intellect* of the animal exhibiting it.

2. If *physical*, it must be the action or stimulus of some physical power or agent employed by the Deity, and under his guidance, so as to work His will upon the organization of the animal, which must be so constructed as to respond to that action in a certain way; or by the exhibition of certain phenomena peculiar to the individual genus or species.

3. If *compound* or *mixed*, it will be subject occasionally to variations from the general law, when the intelligent agent sees fit.

1. With respect to the *first* Hypothesis, one of the principal promulgators and patrons of which is Addison,[1] it nearly amounts to this, as that amiable writer confesses, that "God is the soul of brutes." It is contrary, however, to the general plan of Divine Providence, which usually produces effects indirectly, and by the intervention an l action of means or secondary causes, to suppose that it acts *immediately* upon insects and other animals, and is so intimately connected with them as to direct their instinctive operations; such an action, it should seem, would be infallible, and never at fault, whereas

---

[1] See *Spectator*, ii. p. 121.

observation has proved that animals are sometimes mistaken, where their instinct should
direct them. For, if God were their *immediate*
instructor, would it be possible for the flesh-fly,
as I have seen that she does, to mistake the
blossom of the carrion-plant[1] for a piece of flesh,
and lay her eggs in it; or for a hen to sit upon
a piece of chalk, as they are stated to do,[2] instead of an egg? Still all instincts are from
God, He decreed them, and organized animals
to act according to that decree, and employed
means to impel them to do so.

Other arguments might be adduced proving
that this Hypothesis does not rest upon a sound
foundation; but as I shall hereafter advert to
some of these, I shall now proceed to consider
whether instinct be the action of some *intermediate* intelligence, employed by the Deity,
upon the animal exhibiting it.

An ingenious and acute writer, Mr. French,
is the author of this Hypothesis, which appeared
in the first number of the Zoological Journal.
He infers, "That the Divine Energy does in
reality act, not *immediately*, but *mediately*, or
through the medium of moral and intellectual
influences, upon the nature or consciousness of
the creature, in the production of the various,
and in many instances, truly wonderful actions

---

[1] *Stapelia hirsuta.*          [2] Spectator, ii. n. 120.

which they perform ; that brutes are governed by such agencies, *good* and *evil*, but under the control of Providence ; and that such agencies act by impressions upon their conscious nature, but unperceived by it in a moral or intellectual sense."[1] He thus opens the way to his theory. "If it be asked by what intermediate agency the operations of brutes are thus directed ;—I reply that it is generally admitted by a large class of mankind, at least, that superior (yet intermediate) powers of some kind, are in actual connection with the human mind."[2]

From the passages here quoted, it seems evident (though the author declares that he will not even " venture a suggestion as to the nature of the superior powers here alluded to,")[3] that he had in his mind those good and evil intelligencies that are generally acknowledged to be in actual connection with the human mind ; or, to use the common phraseology, *Angels* and *Demons*. The former being the cause of the *beneficent*, and the latter of the *ferocious* instincts of animals.

When he further observes—"Upon these principles the mixed natures of some animals are satisfactorily explained ;—as in the instance of the *Phoca ursina*, the males of which species manifest the most singular tenderness towards

---

[1] *Zool. Jour.* i. 5, 6.    [2] *Ibid.*    [3] *Ibid.* 6.

their young progeny ; and, at the same time, a savage and persecuting disposition towards their females."[1]

From this passage it would seem that the author was of opinion that the same animal was subject to the agency both of good and evil intermediate intelligences, the one producing its affection, and the other its ferocity.

When our Saviour denominates *serpents* and *scorpions* the power of the *enemy*,[2] it may perhaps be thought that he affords some countenance to this opinion, especially as the evil spirit actually made use of the serpent, as his organ and instrument, when he accomplished the fatal lapse of our first parents from the original rectitude of their nature. But, if we pay due attention to the context, we shall find that, in this passage, as often in other parts of scripture, the symbol is put for the thing symbolized. " *I beheld Satan, as lightning, fall from Heaven,*" says our Lord. " *Behold, I give unto you power to tread on serpents and scorpions, and upon all the power of the enemy.— Nevertheless in this rejoice not that the* spirits *are subject to you.*"[3] The treading therefore on serpents and scorpions was treading upon the *spirits* of which they were figures.

If we duly reflect upon the incongruity of an

---

[1] *Zool. Journ.* i. 7.    [2] *Luke,* x. 19.    [3] *Ibid.* 18—20.

angel and a demon influencing the same animal, in so far as it exhibits instincts partly benevolent and partly ferocious, we shall be convinced that this hypothesis, pursued to all its consequences, cannot stand. Intermediate agents between the Deity and the brute are as much in the place of a soul to the latter, as the Supreme Intelligence would be if his action upon them were immediate, so that the same irrational animal would be alternately a machine impelled by a good or evil intelligence. According to this hypothesis, the bee, that symbol of wisdom, when she sets out upon her beneficent errand of collecting honey and pollen, is acted upon by the *good* angel; but, if she meets with any thing that excites her fear or her anger, she is stimulated to take vengeance upon the object of her displeasure, and to make him feel the puncture of her poisoned dart, by the *evil* one.

This can never be admitted. The same objection too lies against this hypothesis as against the last, that it does not account for the mistakes sometimes made by the animal when endeavouring to accomplish its instinct. It cannot be supposed that, in the case before mentioned, the intelligent intermediate agent would stimulate the flesh-fly to deposit her eggs upon the blossoms of the carrion plant, where the young must inevitably perish from hunger, instead of upon real flesh.

I am next to consider whether instinct be the result of the intellectual powers of the animal itself that exhibits it. If we survey the different tribes of the animal kingdom, we shall find a vast difference between them with respect to intellect. That wonderful pulp, which of all substances is alone able to respond to incorporeal agency: to receive and store up the information collected by the organs of sensation, that it may be ready for future use, and which is the seat of the intellectual faculties, that wonderful pulp appears under very different circumstances in the different Classes of animals; but it has not been made evident that the acuteness of the intellect, though in some instances it seems to do so,[1] depends altogether upon the comparative volume of the brain; for that of the mouse, compared with its size, is greater than that of the half-reasoning elephant.[2] Man indeed, generally speaking, has the largest brain of all animals, but it seems a singular anomaly that persons of very weak intellects have often disproportionately large heads, indicating a great volume of brain. When we leave the vertebrated animals, we find the nervous

---

[1] The brain of the elephant is five times the size of that of the rhinoceros, being as 182 to 35. The space for the brain is smaller in the parrot than in any other bird. *Lit. Gaz. May* 28, 1831. *Philos. Trans.* 1822. 42.

[2] Cuv. *Anat. Comp.* ii. 148.

system, in most, materially altered and degraded, so that more power is given apparently
to instinct and less to intellect. In other animals, as we descend, the nervous system becomes more and more dispersed, so that in those
at the foot of the scale we discern no traces of
intellect, and very few of instinct; and only so
much apparent sensation as is necessary for the
purposes of nutrition and reproduction. I have
made the above observations because they bear
in some degree on the question now before us.
For if we pay due attention to the proceedings of
animals, we shall find that those whose nervous
system is cerebral usually exhibit the most
striking proofs of intellectual action, are most
capable of instruction, and are less remarkable
for the complexity and intenseness of their instincts; while those of the next grade, whose
nervous system is ganglionic, as far as we know
them, though not devoid of intellect, are endued
with a much smaller portion of it, while their
instinctive operations are all but miraculous, and
that where the nervous system is still less concentrated both are greatly weakened, till at the
bottom of the scale they almost disappear.
From hence it seems to follow that extraordinary instinctive powers are not the result of
extraordinary intellectual ones.

But when we reflect further, that even in
cases where the instincts are most complex and

wonderful, the animal practises them *infallibly*, without guide or direction, and is as expert at them when it first emerges into life, as when it has been long engaged in the practice of them; it follows that it must be instructed in them from the first moment of its existence in the state in which it exercises them, by an infallible teacher. The bee, the moment it emerges from the pupa, begins to collect honey and pollen, and to perform all the other manipulations that belong to her instincts.

In the higher animals the case is somewhat different. When they emerge into life, from the womb, or from the egg, it is usually in a state of helplessness, in which at first they can do little or nothing for themselves but suck, or receive food from, their dam. As their organization developes they gradually gain new powers, till they arrive at their acme, or age of puberty.

The young beaver generally remains with its parents till it is three years old, when they couple, and build a cabin for themselves and offspring. The unfledged bird remains quietly in its nest, and is content to receive its food and warmth from its parents, but no sooner are its feathers grown, and its beaked prow and plumy oars and rudder fit it to win its way, in the ocean of air, than, incited by parental exhortations, it makes the attempt, and henceforth is equal to support itself, and to fulfill the biddings of instinct as well as of intellect and appetite. This *storge* stimulates the parent animal while its care of its young is necessary to them and then ceases. This is therefore chiefly instinctive; but in the most intellectual of all animals, where instinctive love ceases, rational love begins; and care and anxiety for the welfare of our offspring, and affectionate regard for their persons, continues after they cease to have any need of our help and attention.

It is not always easy in this tribe of animals to distinguish those actions that are purely instinctive from those that are not so, and writers on this subject, as was before observed, often ascribe to instinct actions that are produced by other causes. Animals of the higher grades, by means of their organs of sensation, acquire ideas upon which they in some sort reason, by comparing one with another; thus they get experience, and as they grow older literally grow wiser. Hence we see old ones often very cunning and expert in removing obstacles, finding their way, and the like.

With regard to truly instinctive actions, they invariably follow the developement of the organization; are neither the result of instruction, nor of observation and experience, but the action of some external agency upon the organization, which is fitted by the Omniscient Creator to respond to its action.

Indeed, if intellect was the sole fountain of those operations usually denominated instinctive, animals, though they sought the same end, would vary more or less in the path they severally took to arrive at it; they would require some instruction and practice before they could be perfect in their operations; the new born bee would not immediately be able to rear a cell, nor know where to go for the materials, till some one of riper experience had directed her. But experience and observation have nothing to do with her proceedings. She feels an indomitable appetite which compels her to take her flight from the hive when the state of the atmosphere is favourable to her purpose. Her organs of sight—which though not gifted with any power of motion, are so situated as to enable her to see whatever passes above, below, and on each side of her—enable her to avoid any obstacles, and to thread her devious way through the numerous and intertwining branches of shrubs and flowers; some other sense directs her to those which contain the precious articles she is in quest of. But though her senses guide her in her flight, and indicate to her where she may most profitably exercise her talent, they must then yield her to the impulse and direction of her instincts, which this happy and industrious little creature plies with indefatigable diligence and energy, till having completed her lading of nectar and ambrosia, she returns to the common habitation of her people, with whom she unites in labours before described,[1] for the general benefit of the community to which she belongs.

More reasons might be adduced to prove that intellect is not the great principle of instinct, but enough seems to have been said to establish that point. It should be borne in mind, however, that though intellect is not the great principle, yet it must be admitted that all animals gifted with the ordinary organs of sensation, more or less employ their intellect in the whole routine of their instinctive operations, as I shall show under another head.

2. But if no metaphysical power can be satisfactorily demonstrated to be the immediate cause of instinct, then it seems to follow that it must be either a physical one, or one partly physical and partly metaphysical.

In the former case, it must be the action of some physical power or agent, employed by the Deity, and under his guidance so as to work his will, upon the organization of the animal ; which must be so constructed as to respond to that action in a certain way, or by the exhibition of certain phenomena peculiar to the individual genus or species.

[1] See above, p. 187, and *Introd. to Ent.* ii. 173.

Mr. Addison has observed—" There is not, in my opinion, any thing more mysterious in nature than this instinct in animals, which thus rises above reason, and falls infinitely short of it. It cannot be accounted for by any properties in matter, and at the same time works after so odd a manner, that one cannot think it the faculty of an intelligent being. For my own part, I look upon it as upon the principle of *Gravitation* in bodies, which is not to be explained by any known qualities inherent in the bodies themselves, nor from any laws of mechanism, but according to the best notions of the greatest philosophers, is an immediate impression from the First Mover, and the Divine Energy acting in the creatures."[1]

I have quoted this passage not as if Addison intended to patronize the hypothesis now before me, but to refer to his illustration of instinct by comparing it with *Gravity*. If Gravity be the result of physical agency, and not an immediate impression of the First Mover, so may Instinct be likewise. Reasoning from analogy it seems inconsistent with the customary method of the Divine proceedings with regard to man, and this visible system of which he is the most important part—for a being that combines in himself matter and spirit, must be more important than

---

[1] Spectator, ii. n. 120.

a whole world that does not combine spirit with matter—to act *immediately* upon any thing but spirit, except by the intermediate agency of some physical though subtile substance, empowered by him to act as his vicegerent in nature, and to execute the law that has received his sanction.

If we consider the effects produced by the great physical powers of the heavens, by whatever name we distinguish them : that they form the instrument by which God maintains the whole universe in order and beauty; produces the cohesion of bodies; regulates and supports the motions, annual and diurnal, of the earth and other planets ; prescribes to some an eccentric orbit, extending, probably, into other systems;[1] causes satellites to attend upon and revolve round their primary planets ; and not only this, but by a kind of conservative energy empowers them to prevent any dislocations in the vast machine; and any destructive aberrations arising from the action of these mighty orbs upon each other. If we consider further what God effects both upon and within every individual sphere and system, throughout the whole universe, by the constant action of those viceregal powers, if I may so call them, that rule under him, whatever name we give them ;

---

[1] *La Place.* E. T. ii. 337. 341.

I say, if we duly consider what these powers actually effect, it will require no great stretch of faith to believe that they may be the *inter-agents* by which the Deity acts upon animal organizations and structures to produce all their varied instincts.

An eminent French zoologist[1] has illustrated the change of instincts, resulting from the modification of the nervous system, which takes place in a butterfly, in the transit to its perfect or imago state from the caterpillar, by a novel and striking simile. He compares the animal to a portable or hand organ, in which, on a cylinder that can be made to revolve, several tunes are noted; turn the cylinder and the tune for which it is set is played; draw it out a notch and it gives a second; and so you may go on till the whole number of tunes noted on it have had their turn. This, happily enough, represents the change which appears to take place in the vertebral chord and its ganglions on the metamorphosis of the caterpillar into the butterfly, and the sequence of new instincts which result from the change. But if we extend the comparison, we may illustrate by it the two spheres of organized beings that we find on our globe, and their several instinctive changes and operations. We may suppose each kingdom of nature to be

[1] Dr. Virey.

represented by a separate cylinder, having noted upon it as many tunes as there are species differing in their respective instincts—for plants may be regarded, in some sense, as having their instincts as well as animals—and that the constant impulse of an invisible agent causes each cylinder to play in a certain order all the tunes noted upon it : this will represent, not unaptly, what takes place, with regard to the developement of instincts, in the vegetable and animal kingdoms ; and our simile will terminate in the enquiry, whose may be that invisible hand that thus shakes the sistrum of Isis,[1] and produces that universal harmony of action, resulting from that due intermixture of concords and discords, according to the will of its Almighty Author, in that infinitely diversified and ever moving sphere of beings which we call *nature*.[2]

What, if the powers lately mentioned, and which, in the Introduction to the present work, I hope I have made it appear, are synonymous with the physical Cherubim of the Holy Scriptures, or the heavens in action which under God govern the universe ; what, if these powers—employed as they are by the Deity so universally to effect his Almighty will in the upholding of the worlds in their stated motions, and prevent-

---

[1] The Sistrum of Isis symbolized the elements.

[2] Φυσις παναιολη.

ing their aberrations,—should also be the intermediate agents, which by their action on plants
and animals produce every physical developement and instinctive operation, unless where
God himself decrees a departure that circumstances may render necessary from any law
that he has established?

With regard to the *vegetable* kingdom, consisting of organized beings without sense or
voluntary motion, few would deny that they are
subject to the dominion of the elements, and
respond to the action of those mysterious powers
that rule, under God, in nature. But when the
query is concerning the *animal* kingdom, most
of the members of which to organization and life
add a will and powers of voluntary motion, and
many have a degree of intelligence residing
within them which governs many of their actions,
we hesitate as to the answer we shall return to it.

It will furnish a presumptive proof that those
actions which are instinctive in animals are the
results of the action of those intermediate powers
to which I have just alluded, if it can be shewn,
that there is any thing in plants at all analogous
to the instincts of animals, for if there be, one
can scarcely suppose that they are produced
by a different cause. Let us, therefore, now
leaving the animal kingdom,—which to us
perhaps appears the sole theatre in which instincts manifest themselves,—and turning our

attention to the vegetable, inquire whether any thing analogous to these springs of action is discoverable there.

One remarkable distinction, between the animal and the vegetable is in the difference of the principles that form their pabulum. The former does not become the nutriment of the latter till it is chemically decomposed ; whereas the latter becomes the food of the former, either in its green, or ripe state, and is not decomposed and turned to nutriment till it is passed into its stomach, and is subject to various actions of various organs, or their products, so that, though the food of both is decomposed in order to be assimilated, yet with regard to the vegetable this happens before it enters it, but to the animal after it enters it, the decomposing powers being without the plant and within the animal. In the former case it is the action of the atmosphere unassisted by the organization of the plant—in the latter it is the same action assisted by the organization of the animal.

Another thing may be here observed—that as the most remarkable instincts of animals are those connected with the propagation of the species, so the analogue of these instincts in plants is the developement of these parts peculiarly connected with the production of the seed —so that the expanded flower and the operations going on in it is the analogue of the reproductive instinct of the animal: this is all produced by physical action upon the organization of the plant. Now if we consider the infinite variety of plants, and the wonderful diversity of their parts of fructification, and that these are all produced in their several seasons and stations by the action of some physical powers upon their varied organization, and by means of the soil in which they are planted, we shall think it nearly as wonderful and unaccountable as the instinctive operations of the various creatures that feed upon them. That the same action should unfold such an infinite variety of forms in one case and instincts in the other is equally astounding and equally difficult to explain.— Compare the sunflower and the hive-bee, the compound flowers of the one, and the aggregate of combs of the other—the receptacle with its seeds, and the combs with the grubs.

Again, as all plants have their appropriate fructification, so they have other peculiarities connected with their situation, nutriment, and mode of life, corresponding in some measure with these instincts that belong to other parts of an animal's economy. Some with a climbing or voluble stem, constantly turn one way, and some as constantly turn another. Thus the hop twines from the left to the right, while the bind-

weed goes from right to left;[1] others close their
leaves in the night, and seem to go to sleep;
others shew a remarkable degree of irritability
when touched; the blossoms of many, as the
sunflower, follow the sun from his rising to his
setting; some blossoms shut up, as in the
anemone, till the sun shines upon them; others
close at a certain hour of the day, as the goats-
beard;[2] another, *Hedysarum gyrans*, slowly re-
volves. The same physical action upon a peculiar organization produces all these effects.

We may further observe that the great majority
of plants send forth radicles which presenting
their points to the sources of vegetable life and
nutrition on all sides, absorb each its portion,
and convey it to the stem from which they
issue; analogous, in this respect, to the polypes,
which unfold and expand their tentacles for a
similar purpose. Ivy planted against a wall or
trunk of a tree supports itself by innumerable
radicles, but I once saw a plant reared as a
standard which sent forth none. This seems
analogous to some animal instincts, which, depending upon circumstances, may be called
*conditional;* as when, in the case of a sterile
queen, the bees do not, as usual, massacre the
drones.[3]

---

[1] See Willd. Princip. of Botany, § 18. n. 51. *a. b.* Plate ii.
*f.* 32, 25.

[2] *Tragopogon.*          [3] *Introd. to Ent.* ii. Lett. xx.

There is another parallelism between the plant and the animal, especially the insect, which appears to prove that their instincts are ruled by the same physical agent, I mean their *hybernation*. In extratropical countries, or a great proportion of them, as the year declines, and the amount of heat, received from its great fountain, is diminished by the shortening of the days, the deciduous trees and shrubs cast their leaves, plants of every description cease more or less their growth, and all vegetable nature seems to become torpid. At the same period, and under the influence of the same cause, the decrease of the amount of caloric, several of the higher animals, all the reptiles, as well as nearly the whole world of insects, retire from the exercise of their wonted instincts, and conceal themselves, some under the earth, and others under bark, under stones, in crevices, moss, and similar hiding places, where they take their winter's sleep, till a more genial temperature whispers to them— *Awake*—and they return to their several employments. This effect in both the plant and the animal, seems to spring from the same *physical* cause—the periodical lowering of the temperature; so that heat appears to be the *plectrum*, and the organization of the animal, the strings it touches, which cause it to exhibit the prescribed sequence of its instincts. Whoever has been in the habit of attending to the motions of *insects* will find them most alert in sultry weather, especially in an electric state of the atmosphere before a thunder storm. Heat and electricity also accelerate the growth of *plants,* if duly supplied with moisture.

It is remarkable, and worthy of particular observation, verifying the old adage that extremes meet, that an approach towards the *maximum* of heat produces sometimes the same effects upon organized nature that an approach towards the *minimum* does. In tropical countries they do not divide the year into winter and summer, but into the rainy and dry seasons; as to temperature, the former would, perhaps, be judged to correspond with our winter, and the latter with our summer, but with respect to the state of animals and vegetables, the reverse would appear to be most consistent with facts. The great rains, according to M. Lacordaire,[1] "begin to fall in Brazil about the middle of September, when all nature seems to awake from its periodical repose; vegetation resumes a more lively tint, and the greater part of plants renew their leaves; the insects begin to reappear: in October the rains are rather more frequent, and with them the insects; but it is not till towards the middle of November, when the rainy season is definitively set in, that all the families appear

---

[1] *Annal. des Sc. Natur.* xx. Juin. 1830. 193.

suddenly to develope themselves ; and this general impulse that all nature seems to receive continues augmenting till the middle of January, when it attains its acme. The forests present then an aspect of movement and life of which our woods in Europe can give no idea. During part of the day we hear a vast and uninterrupted hum, in which the deafening cry of the tree-hopper[1] prevails; and you cannot take a step, or touch a leaf, without putting insects to flight. At 11 a. m. the heat is become insupportable, and all animated nature becomes torpid—the noise diminishes—the insects, and other animals disappear—and are seen no more till the evening. Then, when the atmosphere is again cool, to the matin species succeed others whose office it is to embellish the nights of the torrid zone. I am speaking of the glow-worms[2] and fire-flies ;[3] whilst the former, issuing by myriads from their retreats, overspread the plants and shrubs ; the latter crossing each other in all directions, weave in the air, as it were, a luminous web, the light of which they diminish or augment at pleasure. This brilliant illumination only ceases when the night gives place to the day.

As during our winters, some part of the insect population occasionally appear and dance in the sunbeam, so in Brazil, according to M. Lacor-

---

[1] *Tettigonia. Cicada, &c.*      [2] *Lampyris. Pygolampis.* K.
[3] *Elater noctilucus, &c.*

daire, during the months of May, June, July,
and August, the season of great drought, when
all nature is embrowned, and consequently
affording no proper food for perfect insects ; the
caterpillars of *Lepidoptera* are those mostly to
be met with, while in the rainy season those only
that live in society occur.

The great object of the Creator appears to be
the employment of the various tribes of animals,
to do the work for which he created them at its
proper season ; and where the object is particularly to keep within due limits the growth of
plants, or to remove dead or putrescent substances
before they generate *miasmata*, we may conjecture, that when their services are not wanted,
they would be allowed a season of repose, so
that during winter with us, when there is little
or no vegetation of the plant, and a hot sun does
not cause putrescent substances to exale unwholesome effluvia, the great body of labourers
in these departments, we may say, are sent to
bed for a time, till their labours are again necessary. So also in tropical countries, where
drought and heat united are sufficient to do the
work of nature's pruners and scavengers, by
stopping vegetation, and immediately drying up
animal and other substances, before putridity
takes place, they then abstract themselves, and
retreat to their winter quarters ; but when the
rainy season revives the face of nature, they

return, each to exercise his appointed function, at the bidding of his Creator.

All these circumstances indicate an analogy between certain phenomena observable in the history of *plants,* and some of the instincts of *animals:* and tend to prove that the proximate cause of both may be very nearly related; and that as the immediate cause of the vegetable instinct is clearly *physical,* so may be that of the animal. With regard to all actions, in the latter, which are the result of *intellect,* they, of course, are produced by some principle residing within, as when the senses guide it, or it exercises its memory; and these aid it in following the impulse of instinct. The greatest of modern chemists has observed, with respect to some such agent, " that the immediate connection between the sentient principle and the body may be established by kinds of etherial matter, which can never be evident to the senses, and which may bear the same relation to heat, light, and electricity, that these refined forms or modes of existence bear to the gases."[1] I may observe upon this passage, that the farther any matter is removed from our knowledge and coercion, the more powerful it really is. Thus liquids are more powerful than solids, gases than liquids, imponderable fluids than gases, and so we may

---

[1] *Consolations in Travel,* 214.

keep ascending till we approach the confines of *spirit*, which will lead us to the foot of the throne of the Deity himself, the Spirit of spirits, the only Almighty, the only All-wise, and the only All-good.

Dr. Henry More, a very eminent philosopher and divine of the seventeenth century, under the name of the *Spirit of Nature*, speaks of a power between matter and spirit, which he describes as—" A substance incorporeal, but without sense and animadversion, pervading the whole matter of the universe, and exercising a plastical power therein, according to the sundry predispositions and occasions in the parts it works upon, raising such phenomena in the world, by directing the parts of matter and their motion, as cannot be resolved into mere mechanical powers—which goes through and assists all corporeal beings, and is the vicarious power of God upon the universal matter of the world. This suggests to the *spider* the fancy of spinning and weaving her web; and to the *bee* of the framing of her honey-comb; and especially to the *silk-worm* of conglomerating her both funeral and natal clue; and to the *birds* of building their nests, and of their so diligent hatching their eggs."[1]

This Spirit of Nature of Dr. More seems not very different from the Etherial Matter of Sir

---

[1] *On the Immortality of the Soul*, B. iii. c. 12, 13.

H. Davy; and it is singular, that Dr. Paris, in his interesting life of our great chemist—speaking of a monument to be erected to his memory at Penzance—should thus express himself. " It was to be erected on one of those elevated spots of silence and solitude where he delighted, in his boyish days, to commune with the elements, and where the *Spirit of Nature* moulded his genius in one of her wildest moods."[1]

But—to return from this digression to Sir H. Davy's etherial matter bearing the same relation to heat, light, and electricity, that they do to the gases—I would ask, if such may be the powers by which the soul moves the body, and produces those actions that are in our own power to do or not to do, depending upon the will, does it seem incongruous that light, heat, and air, or any modification of them, upon which every animal depends for life and breath, and nutrition and growth, and all things, should be employed by the Deity to excite and direct them, where their intellect cannot, in their instinctive operations? That their organization, as to their instruments of manducation, motion, manipulation, &c. has a reference to their instincts every one owns; can we not, therefore, conceive that the organization of the brain and nervous system may be so varied and formed by the Creator, as to respond,

---

[1] *Life of Sir H. Davy*, 4to. edit. 517.

in the way that he wills, to pulses upon them from the physical powers of nature; so as to excite animals to certain operations for which they were evidently constructed, in a way analogous to the excitement of appetite? The new-born babe has no other teacher to tell it that its mother's breast will supply it with its proper nutriment; it cries for it; it spontaneously applies its mouth to it; and presses it under the bidding of appetite resulting from its organization. When it arrives at the age of dentition, it as naturally uses its teeth for mastication; it wants no instructor to inform it how they are to be employed to effect that purpose; and so with respect to other appetites which the further developement of its organs produces.

It may, perhaps, be urged, in the case lately alluded to, of the infant growing up to puberty, that the instinctive operations that take place under the bidding of appetite fall under the general law of instinct; but it must be admitted that the gradual developement of the organization is the consequence of the action of physical powers in the processes going on in the body. Or, as a learned writer on the subject asks,—" In effect is instinct any thing else, but the manifestation without of that same wisdom which directs, in the interior of our body, all our vital functions."[1]

[1] Dr. Virey. *N. D. D'Hist. Nat.* xvi. 293.

Having rendered it probable that those instincts, which result evidently from what are called *bodily* appetites, are the consequences merely of physical action upon an organization adapted to respond to it, I shall next inquire whether this may not be the case in instances which are not to be regarded in that light.

We may divide instincts into *three* general heads:—

a. Those relating to the multiplication of the species, especially the care of animals for their young both before and after birth.

β. Those relating to their food.

γ. Those relating to their Hybernation.

a. The pairing of animals usually begins to take place in the spring, when the winter is passed, the earth is covered with verdure and adorned by the various flowers that now expand their blossoms, in proportion as the great centre of light and heat more and more manifests his power over the earth ; the birds sing their love-songs ; the nightingale is now—" Most musical, most melancholy ;"—the cuckoo repeats his monotonous note; and every other animal seems to partake of the universal joy. All this appears the result of a *physical* rather than a *metaphysical* excitement.

As to their care of their future progeny, a great variety of circumstances take place. Vivi-

parous animals have generally to give suck to
their young for a time; oviparous ones either to
construct a nest to receive their eggs, and, after
hatching, to provide them with appropriate food
during a certain period, or to deposit their eggs
where their young progeny, as soon as hatched,
may infallibly find it. But first, I must say
something of that *Storge*, or instinctive affection,
which is almost universally exhibited by females
for their progeny both before and after parturition; a feeling of affection not generally
common to the males, or rather only in a few
instances, as where the male bird assists the
female in incubation. Yet this instinctive fond-
ness, as soon as it ceases to be necessary,
vanishes; except, as was before observed,[1] in the
human species; a fact that seems to prove that
it is not the result of the association of ideas,
but of an impress of the Creator interwoven with
the frame. But that this impress is by means
of a physical interagent, seems to follow from
this circumstance—that the *hen* shows the same
instinctive attachment to the young *ducklings*
that have been hatched under her, that she
would do to chickens, the produce of her own
eggs; and if the new-born offspring of any
mammiferous animal is abstracted from her, and
another substituted, even of a *different* kind,

[1] See above, p. 238.

the same affectionate tenderness is manifested towards it, as its own real offspring would have experienced. Now was it a metaphysical, and not a physical, impulse, surely this would not be the case. This is only one of many instances, which prove that instinct is not infallible: and, in truth, with regard to the higher animals, many associations may take place between the child and parent that help to endear the former to the latter. In the first place, the very circumstance of its being the fruit of her own bowels, and fed with milk from her own breast must bind it to her by the tenderest of ties; especially as, at the same time, it relieves her from what is troublesome. There is something also in infant helplessness, and infant gambols, calculated to win upon the doting mother. The subsequent alienation and estrangement of the female from her young, which takes place in all animals except man, appears, in the first instance, to be produced by their becoming troublesome and annoying to her; which, in some degree, may account for her desire to cast them off. Examining the subject, therefore, on all sides, in the highest grades of animals, and those in whom maternal affection appears most intense, intellect and associations may be a good deal mixed with instinct in producing it. As we descend in the scale, the intensity of the feeling seems much reduced; and, in numerous tribes, is confined solely to the circumstances of parturition. So that the *Storge*, and its cessation, do not appear altogether so extraordinary and unaccountable as a cursory view might tend to persuade us.

The *Mammalians*, in general, appear to have recourse to very few striking preparatory actions previously to bringing forth their young, since they have usually no nest to prepare for their reception. Cats, however, it may be observed, search about very inquisitively for a snug and concealed station; and burrowing animals naturally retire to the bottom of their burrows, when their feelings tell them their hour is come, and there are relieved of their precious burthen. Several others of the *Rodentia*, or gnawers, as the dormouse, make beds of their own hair to receive their young. In most cases that fall under our daily observation, the young are dropped where the mother happens to be when the pains of labour overtake her. The animals we are speaking of have at hand immediately a plentiful supply of food for the nutriment of their new-born offspring; they have not, like the birds, to search for provision for them, but, from their own bodies, furnish them with a delicious fluid suited to their state, which forms their support till they are able to crop and digest the herbage, when they are left to shift for themselves. Some are born more independent of maternal care than others; thus domestic animals, as the calf, the lamb, and the young colt, can move about almost as soon as they are born, and can immediately use their organs of sight; whereas the progeny of beasts of prey usually come into the world blind, and some time elapses before they can run about, so that the dam, if she wishes to remove them, must carry them herself, which she generally does, in her mouth.

As the proper food of herbivorous quadrupeds is almost every where abundant, they are soon tempted, without the intervention of the mother, to browse upon the herbage : but the predaceous beast whose food must be pursued and captured, takes more pains to instruct her young how to maintain themselves; thus the cat lays the mouse or bird, that she has caught, before her kittens; and it is laughable to observe how they are excited, and with what resolution and ferocity the little furies endeavour to keep possession of the prey their dam has brought to them.

But of all classes of animals the *birds* are the most remarkable for the labours they undergo preparatory to laying their eggs. In those that migrate a long aërial voyage is previously to be undertaken, the stimulus to which, in the swallow, appears to be altogether physical,[1] and is

---

[1] Vol. 1. p. 102. See Jenner, *Philos. Trans.* 1824, 20.

probably so in other migrators.   But what is it
that directs them in their flight, and enables
them to return to the countries from which
they had migrated?   Did the swallow[1] steer her
course within sight of land, it might, perhaps,
be supposed that her *memory* was her director :
but these birds are often found at sea, hundreds
of miles from any shore,[2] where, one would
think, there could be no index either in the
clouds or the ocean to instruct her which way to
steer her adventurous course.   The only atmospheric phenomenon affecting her would be the
change of temperature as she went northward.
But we can only conjecture in this case—observation, as well as scripture, tells us, indeed,
*The stork in the heaven knoweth her appointed*
*times; and the turtle, and the crane, and the*
*swallow observe the time of their coming*,[3] but,
God, who decrees the end, appoints the means,
which often remain amongst his *Secret Things.*
Yet, though the immediate agent that guides
the swallow over the expanse of water, from the
torrid to the temperate zone is latent, we may
still inquire, when she has made the shores of
Britain, what is it that urges her to seek her old
vicinity, and to build her nest in the very spot
where she herself first drew breath, as Dr. Jenner's experiments prove that swallows do?[4]

---

[1] *Hirundo rustica.*          [2] *Philos. Trans.* ubi supr. 13.
[3] *Jerem.* viii. 7.           [4] *Philos. Trans.* ubi supr. 16.

Here may we not conjecture that her intellect and memory become her guides? She recognizes the spot in which she committed herself to the sea breeze; and there, probably, again flies inland, and will have no great difficulty in pursuing the line of country which leads to her native village, and to the very roof under the eaves of which she was born.

But of all the instincts of the feathered part of the creation, there is none more remarkable, more varied, and more worthy of admiration than that which directs them in the situation and structure of their nests.—One nidificates upon the ground;[1] another under ground, or in the sand;[2] some select the chimney or eaves of houses for their clay-built structures;[3] those gelatinous nests, which the Chinese epicures and orators so highly prize, are formed in caverns and dark places by the little bird[4] whose work they are. The great majority, however, nidificate in trees and bushes, and where they are within reach their nests are carefully concealed.

The structure and materials of nests are also infinitely various, and may be considered to result, as well as all the proceedings of animals with regard to their young, from an excitement analogous to that which Dr. Jenner first noticed

[1] *Motacilla Troglodytes.*     [2] *Hirundo riparia.*
[3] *H. rustica et urbica.*     [4] *H. esculenta.*

in the swallow;[1] upon which he observes—
" The economy of the animal seems to be regulated by some *external* impulse which leads to
a train of consequences,"[2] and which does not
cease its action till it has accomplished the end
for which it was given; namely, the procreation;
oviposition preceded by nidification; incubation;
hatching, or birth; nutrition and education of
the young progeny of each individual kind,
according to the general law of the Creator.

We know very little of the proceedings of the
remaining Classes of Vertebrates—which are
distinguished by having cold blood—the *Reptiles*, namely, and the *Fishes;* except that they
do not feel that instinctive love for their young,
after birth, exhibited by the quadrupeds and
birds. They, however, are invariably instructed
by the Creator to select a proper place in which
to deposit their eggs where they can be hatched
either by artificial or solar heat. Those of some
*Ophidians*, as snakes, are buried in sand, and
not seldom even in heaps of fermenting manure;
while those of venomous ones are hatched in the
womb of the dam, and come forth in the serpentine form. The *Saurians* also select a proper place for their eggs, and then desert them;
the crocodile buries hers in the sands near the
river; where many, however, are devoured by

---

[1] *Philos. Trans.* 1824. 20.    [2] *Ibid.* 25.

the ichneumon, and its other enemies, and are even relished by man. In the *Batrachian* Order one species of *salamander*[1] commits a single egg to a leaf of the *Persicaria*, which it protects by carefully doubling the leaf, and then, proceeding to another, repeats the same manœuvre, till her oviposition is finished :[2] the *toads* and *frogs* lay their eggs in the water, the former producing two long strings resembling necklaces, formed, as it were, of beads of jet, inclosed in crystal; while those of the latter consist of irregular masses of similar beads. This gelatinous or transparent envelope forms the first nutriment of the embryo. The nuptial song of the Reptiles is not, like that of birds, the delight of every heart, but is rather calculated to disturb and horrify than to still the soul. The hiss of serpents; the croaking of frogs and toads; the moaning of turtles; the bellowing of crocodiles and alligators,[3] form their gamut of discords.

With regard to the Class of *Fishes*, the general object of those that migrate appears to be the casting of their spawn; this it is that causes the different species of the *salmon* genus to leave the sea for the rivers; for this the *herring* travels southward, and the *mackarel* seeks the north; all of them guided by the law of the Most High, shewing itself by an indomitable instinct, to

---

[1] *Salamandra platycauda.*
[2] *Edinb. Phil. Journ.* ix. 110.      [3] See Vol. I. p. 32.

seek those stations for oviposition that are best suited to the aëration, hatching, and rearing of their spawn;—but as no very striking traits are upon record with regard to the oviposition of fishes, I shall merely refer the reader, with respect to the instinct of the migrators, to a former part of the present work, where that subject is discussed more at large.[1]

Under this head I shall only further notice the numerous tribes of the *insect* world, which have all their seasons, varying according to their several destinies, for fulfilling the great law of nature, and to which the organization of each species is adapted : and when the period for laying their eggs is arrived each is directed to place them where their young, when disclosed, may find their appropriate nutriment. From the instance of the flesh-fly, above related,[2] we learn that it is their *scent* that directs insects to a proper station for their eggs. When we re-collect that every plant, almost, is the destined food of some peculiar insect, we may conjecture that the sense of smelling must, in them, be far more nice than in the higher animals, so as to enable them to distinguish from all others the appropriate nutriment of their own descendents. Where the parent, as is sometimes the case, feeds upon the same plant with the children,

---

[1] See Vol. I. p. 107.    [2] See above, p. 231.

she requires no such guide, but with respect to the majority of insects, especially the infinite host of *Lepidoptera*,—which, after they arrive at their perfect state, never touch what forms their nutriment while they are larves,—some such guide is absolutely necessary.

β. Another Class of. Instincts relates to the different modes by which animals procure their *food*. Nothing affords a more striking proof of Creative Wisdom, and of the most wonderful adaptation of means to an end, than the diversities of structure with a view to this particular function. If we consider the infinite variety of substances, animal and vegetable, produced from the earth, which form the nutriment of its inhabitants—some solid and not easily penetrable ; others soft and readily severed and comminuted ; others again fluid, or semi-fluid ;—we may conceive what a vast diversity of organs is necessary to effect this purpose. To render solid food, of any kind, fit for deglutition and digestion, the same mouth must be furnished with several kinds of teeth, some for incision, others for laceration, others again for grinding and mastication—while those that only absorb liquids merely require an organ adapted for suction, though often, at the same time, fitted to pierce the substance from which the nutritive fluid is to be derived. How various, also, must be the organs

for swallowing, and digesting the food according
to its nature; others for elaborating it, and
abstracting from it all those substances that are
required by the several systems at work in the
body, and conveying them to their proper stations; and the means also for rejecting from the
body the residuum after the secernment for the
above purposes of the finer life-supporting products. Here are a variety of organs, admirable
in their structure, and fitted for action in an
infinity of ways; some at the bidding of the will
stimulated by the appetite; others independent
of the will, such are the distillations, percola-
tions, chemical and electrical processes, constantly going on in the body of every animal,
to separate all the products that its nature and
functions require, all speak of a *mechanical*
agency at work within, not independent in its
operation, but fulfilling a law which must be
obeyed.[1] It has been found that *Galvanic action*
will supply the place of the *will* upon the nerves
and muscles, for by it the eyes can be opened,
and other muscular movements be produced in
a dead body.[2] Sir H. Davy was of opinion that
the air inspired carries with it into the blood a
subtile or ethereal part probably producing
animal heat, since those animals that possess

[1] See Dr. Roget's excellent statements on these subjects,
*B. T.* ii. chap. iii.—ix.

[2] See Dr. Wilson Philip in *Philos. Trans.* 1829. 271, 278.

the highest temperature consume the greatest quantity of air, and those, that consume the smallest quantity, are cold blooded.[1]

The herbivorous *Mammalians* are generally not remarkable for any *artificial* means of procuring their food. Providence has spread a table before them, and invites them to partake of it, without any other trouble, than bending their necks to eat it; but the carnivorous ones, —as their destined pabulum is endued with locomotive powers, which enable it often to escape from them, and disappoint their expectations,—must have recourse to stratagems, and lie in wait for their prey; these, however, consist chiefly in concealing themselves and springing suddenly upon it. The fox, of all quadrupeds, is the most celebrated for his stratagems and finesse in entrapping his game, and his patience is equal to his craft. Some have doubted whether this animal can *fascinate* poultry, as has been often asserted, but I know one instance which fully confirms it. A friend of mine one night hearing a noise, upon looking out in its direction, saw a fox under the hen-roost, peering up at the hens, which both he and his wife, who told me the story, saw, as they did also the fox running away, in spite of their shouting, with one in his mouth. Indeed, on

[1] *Consolations in Travel,* 196, 197.

any other principle we cannot account for his depopulating the hen-roosts in the night.

The *birds* are less noted, than even the quad-rupeds, for their stratagems, or any remarkable means of providing food for themselves or their young. Those of prey boldly attack and seize their destined food wherever they find it; the owls, indeed, like the cats, their analogues, seem to use artifice as much as strength to attract the mice. The carrion-feeders, as the vultures and crows, soon discover the carcasses of dead animals.[1] Some of the sea-birds, espe-cially the gulls, indicate the approach of bad weather, by leaving the coast, and seeking the interior; and, during the intense frosts of a severe winter, the web-footed birds and waders, quitting their summer stations in the more northern regions, fly to the south and seek the unfrozen springs and waters of the inland dis-tricts, where they find a supply of food. All these physical actions seem to arise from a phy-sical cause, and easily to be accounted for, with-out having recourse to any other.

With regard to the cold-blooded animals, the fishes and reptiles, we know but little of their habits in this respect, or of any particular stratagems to which they have recourse to pro-cure their food. Some of the predaceous fishes, as the pike and perch, appear to lie in wait in

[1] Roget, *B. T.* ii. 407.

deep water, and so dart upon their prey; others, as the shark, with open mouth pursue and devour them; the fly-catching ones, as the several species of the carp and salmon genus,[1] are equally upon the watch, but nearer the surface, to seize a may-fly[2] or ephemera; the fishing-frog[3] hangs out its lines in the sea to catch other fishes; the serpents are said to fascinate the birds; the enormous boa lies in wait for the antelopes and other quadrupeds, and coiling itself round them in mighty folds, crushes them to render them more fit for deglutition; the Batrachians, Chelonians, and numerous Saurians are on the alert after insects and small game; while the vast and ferocious crocodiles and alligators, looking like trunks of trees, lie basking near the surface of the water, ready to spring upon any large fish, or even man, that may chance to come within reach.

Of all animals, *insects* afford the most numerous instances of instinctive proceedings with this sole end in view; the pit-falls of the ant-lion;[4] the webs and nets of the various sorts of spiders spread over the face of nature; and many more, furnish instances of stratagems to secure their daily food; while an infinity of others acquire it, aided only by their senses and natural weapons. Let any one look at the pro-

---

[1] *Cyprinus* and *Salmo.*     [2] *Phryganea.*
[3] *Lophius.*     [4] *Myrmeleon.*

minent eyes, tremendous jaws, and legs and
wings formed for rapid motion on the earth or in
the air of the tiger-beetles,[1] and he will readily
see that they want no other aid to enable them
to seize their less gifted prey : and numerous
other tribes both on the earth and in the water
emulate them in these respects.  The *pacific* or
herbivorous insects also are mostly fitted with
an extraordinary acuteness of certain senses to
direct them to their appropriate pabulum.  The
sight of the butterfly and moth invariably leads
them to the flowers, to suck whose nectar
their multivalve tubes are given them.  The
scent of the dung-beetles and the carrion-flies
allures them to their respective useful, though
disgusting, repasts.  A very numerous tribe of
those that derive their nutriment from other
animals, neither entrap them by stratagem, nor
assail them by violence; but, as the butterfly
and the moth deposit their eggs upon their
appropriate *vegetable*, so do these upon their
appropriate *animal* food.  Every bird almost
that darts through the air, every beast that
walks the earth, every fish that swims in its
waters, and almost all the lower animals, and
even man himself, the lord of all, are infested in
this way.

Upon the food of the *Crustaceans, Molluscans,*
and all the lower grades of animals, I have

[1] *Cicindela.*

before sufficiently enlarged ; I need not, therefore, here resume the subject.

Thus we see the Almighty and All-wise manifests his *goodness*, as well as his wisdom and
power, in providing for the wants of all the
creatures that he has made; fitting each with
peculiar organs adapted to its assigned kind of
food, both for procuring it, preparing it, digesting it, assimilating it, and for rejecting the
residuum of all these operations. A physical
action upon each of these organs and systems,
fitted by him to receive and respond to it, is all
that the case seems to require in the majority of
instances : in those, however, that depend upon
artifice and stratagem for their food, the exciting
cause is less obvious. These, indeed, belong to
the higher instincts considered under the *first*
head.

γ. That class of Instincts which relates to the
*hybernation* of animals having been considered
in another place,[1] I shall only observe here, that
the action of a physical cause is in no department of the history of animals more evidently made out.

My learned friend and coadjutor, Mr. Spence,
has, in the *Introduction to Entomology*, produced several facts, as not easily reconcilable
to the hypothesis with respect to the cause of

---

[1] See above, p. 248.

Instinct which I am now considering; and probably a great many more might be brought forward; but my object here is merely to consider the general principle; it would, indeed, be needless and endless to discuss particular cases, and fully to account for all aberrations, which, in the present state of our knowledge, it would not be possible to do.

But there is one circumstance of a less confined nature, and upon which a good deal of the question hinges, to which it will be proper to advert. I mean the change that has been observed in the nervous system of some insects in their passage from one state to another. It is contended that this change has nothing to do with any alterations that then take place in their instincts, but only with those in their organs of sense or motion.[1] In confirmation of this opinion it is further affirmed, that in three whole Orders,[2] the structure of the nervous chord is not altered, and yet they acquire new instincts.

But though no change has been *noticed* to take place in the number of ganglions of these Orders, there must necessarily be a developement in those that render nerves to the wings and reproductive organs; so that, though some ganglions may not become confluent, as in the *Lepidoptera*, yet the range of their nerves

[1] *Introd. to Ent.* iv. 27, 28.
[2] Viz. *Orthoptera, Hemiptera,* and *Neuroptera.*

is increased. In this respect, they are in much the same situation with the higher animals, though their nervous system, as to its organization, undergoes no material change, yet from the period of their birth, it is gradually more and more developed till they arrive at the age of puberty, when new appetites are experienced and new powers acquired, not by *metaphysical*, but by *physical*, action upon their several systems. In the three Orders referred to by Mr. Spence, there is not that difference between the different states of the insects that compose the majority of them, that there is between those whose pupes are not locomotive. The larves of the locust, for instance, are stated to emigrate, as well as the perfect insect, and live upon the same food ; the only difference is in the locomotive and reproductive powers of the latter, both of which, as I have just said, must be connected with some change in their nervous system, operated gradually by a physical agent.

From what has been stated, with respect to these several classes of instincts, it appears, that, as far as can be judged from circumstances, they have their beginning in consequence of the action of an intermediate physical cause upon the organization of the animal, which certainly renders it extremely probable that such is the general proximate cause of the phenomena in question. I would, however, by no means, be understood to assert this dogmatically, but merely that it appears to me the most probable hypothesis, and most consistent with the analogy of the Divine proceedings in this globe of ours, as well as with his general government of the heavenly bodies ; and though I have mentioned heat, electricity, and other elements as concerned in the production of these phenomena, yet I do not assert that other physical principles may not be commissioned to have a share in it. This field is open both to the speculatist and experimenter ; they may each assist the other in traversing and exploring it, and the well known adage, *Dies diem docet*, be verified more and more by their united efforts.

Some may still feel disposed to ask,—Is it within the sphere of probability, or even possibility, that by the mere action of physical powers, however subtile, upon the brain and nerves of an animal there should be produced such a wonderful sequence of actions and manipulations as we know to be exhibited by the *beaver*, the *bee*, the *spider*, and the *ant ?* Actions confessedly above the range of their intellect. But to this I would answer, we know that with God all things are possible that do not imply a contradiction ; and His Wisdom, Power, and Goodness, may be as evidently, and more evidently, manifiested, by the infinite varieties in the organization necessary to excite the appetite for such and such instinctive employments and operations; and to stimulate animals always to run the same prescribed routine of action from day to day, and year to year; than if he did it by his *own* immediate action upon them, or that of his *ministering*, or *other*, *spirits*.

When we examine a time-piece contrived by a skilful artist, containing within it various wheels and other movements, all acted upon by one main spring or pendulum; by means of which, influencing all, seconds, minutes, and hours are indicated as they pass; and the latter are struck successively, and repeated if required: we admire the work, but more the art and hand that contrived and executed it; but our admiration would be much diminished, if, instead of these effects being produced by the action of a main spring or pendulum upon its organization, if I may so call it, it was necessary that the maker of the machine, or one of his operatives, should always be present to move the hands or strike the hours. So it seems most to magnify the Power and Wisdom of the Creator, if we suppose him to act by physical means in all cases above the intellect of the animal. If he governs the physical universe by such means, is it much to suppose, that by the same he moves a bird, or a bee, to glorify him by their admirable instincts? Where action is indeed from the Deity *upon spirit*, as upon the soul of man, in a certain sense, it is *by spirit;* either immediately as by the Holy spirit; or mediately as by an angelic nature; but *below* spirit, it is surely most consonant to every thing that we see and know, that it should be by an agent below spirit.

3. I am now arrived at the last supposition or hypothesis—that the cause instinct may be *compound* or *mixed*—in some respects physical, in others metaphysical. In this case it will be subject occasionally to variations from the general law when the intelligent agent sees fit.

But upon this head I shall not be very long, and I only introduce it here, to shew that the Deity sometimes dispenses with the general law of instinct, or permits it occasionally to be interfered with by the will of the animal, or other agency. All animals that exercise instinctive operations, have in their several organs of sensation, certain guides given to enable them to fulfil those instincts so as to bring about the purposes of Providence.

Sight, hearing, scent, taste, touch, perception, influence the will, and direct each animal to the points in which its instinctive actions are to commence; and so far instinct is, as it were, *mixed* with intellect. I have seen it somewhere observed—that instinct in conjunction with a

principle of limitation,—*the intellectual faculties,*
—rules the actions of all *sentient and organised*
beings; just as gravity with the principle of
counteraction—*repulsion*—determines the place
and composition of all *inorganic* bodies.

With regard to the Deity, he retains in his
hands the power of suspending or altering the
action of the laws that have received his sanction; and permits other metaphysical essences
to do the same. When females overcome that
*storge* or instinctive love for their offspring,
either from the dread of shame, or worse motives,
and destroy them, in common parlance, we say
that they were tempted by an *evil spirit* to commit the crime. Mr. Bennet, in his interesting
*Wanderings in New South Wales, &c.,* relates
that it is common for the females of the abo-
riginal tribes, if they experience much suffering
in their labour, to threaten the life of the poor
infant, which when born they barbarously destroy.[1] This is a fearful counteraction of instinct
flowing from an *evil* source.

The Deity himself, doubtless when there is
—*Dignus vindice nodus*—sometimes suspends the
action of an instinct. It is related in the Holy
Scripture, that when the ark of God was taken
by the Philistines, in order to ascertain whether
the plagues that were sent upon them were from

---

[1] I. 122.

God, they yoked two milch kine that had calves
to the cart in which it was sent to Bethshemesh,
and the kine went straight to that place, their
instinct being mastered by a strong hand, though
they went lowing after their calves all the way.[1]
Here the Deity ruled the instinct. God inter-
feres with the instincts of animals also when he
prescribes their course and sends them in any
particular direction to answer his purpose: as in
the case of the prophet Jonah.[2] Properly speaking, those interpositions of the Deity by which
the law of instinct is suspended, to answer a
particular purpose of his Providence, like that
just related, must be regarded as miraculous;
but yet, though unrecorded, they may happen
oftener than we are aware in the course of his
*moral* government; sometimes perhaps also to
remedy some *physical* evil. This appeared therefore a proper place to advert to them.

[1] 1 *Sam.* vi. 7. 12.　　　　[2] Vol. I. p. 263.

## Chapter XIX.

*Functions and Instincts. Arachnidan, Pseuda-rachnidan, and Acaridan Condylopes.*

HAVING wandered long enough, perhaps too long, in a wide and mazy field, but fertile everywhere in proofs of the Power, Wisdom, and Goodness of the Creator, it is time to return to the high road from which we diverged.

The Class of animals which led me into this digression were the Myriapods, concerning which I observed, when I commenced my account of them, that on quitting the Crustaceans, the way seemed to branch off from the long-tailed Decapods by them, and from the short-tailed ones by the Arachnidans. We are now then to give a history of the latter Class.

Latreille, in which he has been followed by most modern Arachnologists, in his work in aid of Cuvier's last edition of the *Règne Animal*,[1] divides his Arachnidans into two Orders, *Pulmonaries*, or those that breathe by *gills*, and *Trachearies*, or those that breathe by *spiracles* in connection with *tracheæ*. In his latest work,[2]

[1] *Les Crustacés, les Arachnides, et les Insectes.*
[2] *Cours D'Entomologie.*

which he did not live to finish, he added a third Order, including some parasites, infesting marine animals, such as the whale louse.[1] These, from their having no apparent respiratory apparatus, he named, *Aporobranchians*.

As the pulmonary *Arachnidans* of Latreille differ from the Trachearies, &c., not only in having their body divided into two sections, but likewise both in their respiratory organs and those of circulation, I have always regarded them as forming a distinct Class.[2]

The following characters distinguish this Class:

BODY covered by a coriaceous or horny integument, divided into two segments. *Head* and *trunk* confluent so as to form a single segment, denominated the *Cephalothorax*. *Eyes*, 6—8. *Legs*, 8. *Spinal chord*, knotty. A *heart* and *vessels* for circulation. *Respiration* by *gills*. *Sexual organs*, double.

This Class consists of two Orders.

1. *Araneidans.* *Integument* coriaceous. *Mandibles*, also called *cheliceres*, consisting of a single joint, armed with a claw, perforated near the apex for the transmission of venom, and when unemployed folding upon the end of the mandible. *Gills*, 2—4. *Abdomen*

---

[1] *Nymphon grossipes.*    [2] *Introd. to Ent.* iii. 19. 24.

united to the trunk by a foot-stalk. *Anus* furnished with 4—6 spinning organs.

2. *Pedipalps.*[1] *Integument* horny. *Feelers* extended before the head, armed with a forceps or didactyle claw. *Abdomen* sessile. *Gills*, 4—8.

1. *Araneidans*, or spiders.

No animals fall more universally under observation than the *spiders;* we see them everywhere, fabricating their snares or lying in wait for their prey, in our houses, in the fields, on the trees, shrubs, flowers, grass, and in the earth; and, if we watch their proceedings, we may sometimes see them, without the aid of wings, ascend into the air, where, carried by their web as by an air-balloon, they can elevate themselves to a great height. The webs they spin and weave are also equally dispersed; they often fill the air, so as to be troublesome to us, and cover the earth. M. Mendo Trigozo[2] relates, that at Lisbon, on the 6th of November, 1811, the Tagus was covered, for more than half an hour, by these webs, and that innumerable spiders accompanied them which swam on the surface of the water. I have in another place[3]

---

[1] *Manipalps* would be a more proper term, as the feelers are used for prehension, not for walking.

[2] Latr. *Cours. D' Ent.* i. 497.     [3] See above, p. 183.

given an account of the instruments by which they weave them ; and shall now say a few words upon those by which their Creator has enabled them to produce the material of which they are formed.

At the posterior extremity of the abdomen, formed usually by a prominence, is the anus, immediately below which, planted in a roundish depressed space, are four or six jointed teat-like organs, of a rather conical or cylindrical shape. The exterior pair is the longest, consisting of three joints ; but these have no orifices at their extremity for the transmission of threads : the other four[1] consist each of two joints, and are pierced at their extremity with innumerable little orifices, in some species amounting to a thousand from each, from which their web issues at their will, or bristled with an army of infinitely minute biarticulate spinnerets,[2] each furnishing a thread at their extremity. These teats are connected with internal reservoirs, which yield the fluid matter forming the thread or web. These reservoirs in some species consist of *four*, and in others of *six* vessels folded several times, and communicating with other vessels in which the material that forms their web is first elaborated.[3]

---

[1] *Mammulæ, Introd. to Ent.* iii. 391.    [2] *Fusi, ibid.* 392.

[3] Latr. *Cours D'Ent.* i. 496.

Such are the organs which furnish the material of those wonderful and diversified toils which the spiders weave to entrap the animals that form their food.

The threads, after they issue from these organs, are united, or kept separate, according to the will or wants of the animal; and it is stated, that from them certain spiders can spin *three* kinds of silk.[1] Their ordinary thread is so fine, that it would require twenty-four united to equal the thickness of that of the silkworm. These threads, fine as they are, will bear, without breaking, a weight sextuple that of the spider that spins them. They employ their web, generally, for three different purposes; in the construction of their snares, of their own habitations, and of a cocoon to contain their eggs.

Spiders were divided by the older Arachnologists, after Lister, into families according to the mode in which they entrap or seize their prey. More modern writers[2] on the subject, have taken their respiratory organs as regulating the primary division of the Order: upon this principle, the spiders are formed into two tribes, those that have two pairs of gills;[3] and those that have only one pair.[4] M. Walckenaer, who

[1] Blackwall, in *Linn. Trans.* xvi. 479.
[2] L. Du Four. Latreille.
[3] *Tetrapneumones.* Latr. *Theraphosa,* &c. Walck.
[4] *Dipneumones.* Latr. *Aranea.* Walck. excluding *Dysdera.*

has studied the Order more than any man in Europe, has not only divided the above two tribes into genera, &c., from characters taken from their form and organization; but has also considered them with respect to their habits, and under this head, divides them into four sections:

1. *Hunters*, wandering incessantly to entrap their prey.
2. *Vagrants*, watching their prey, concealed or inclosed in a nest, but often running with agility.
3. *Sedentaries*, forming a web in which they remain immoveable.
4. *Swimmers*, swimming in the water to catch their prey, and there forming a web.

To the first tribe, those, namely, with *four* gills, some spiders belong, the instincts of which are very remarkable. One of the largest, and most celebrated, is the bird-spider.[1] It forms the tube which it inhabits of a white silk like muslin, which it fixes amongst leaves, and in any cavities, and there watches its prey; it is accused by some of destroying even birds, whence its name, especially the humming-bird:[2] but this rests upon questionable authority; and writers are not agreed as to its general habits. Probably several species are confounded under

[1] *Mygale avicularia.*        [2] *Trochilus.*

the same name. I shall not therefore enlarge further on its history; I mention it merely as the largest spider known.

The proceedings of those called the *trap-door* spiders[1] are better authenticated, as those of the mason-spider by the Abbé Sauvages,[2] and those of another species very recently, in the annals of the French Entomological Society, by M. V. Audoin, one of the most eminent of modern entomologists, under the name of the *pioneer*;[3] of his interesting memoir, I shall here give a brief abstract.

Some species of spiders, M. Audoin remarks, are gifted with a particular talent for building: they hollow out dens; they bore galleries; they elevate vaults; they build, as it were, subterranean bridges: they construct also entrances to their habitations, and adapt doors to them, which want nothing but bolts, for without any exaggeration, they work upon a hinge, and are fitted to a frame.[4]

The interior of these habitations, he continues, is not less remarkable for the extreme neatness which reigns there; whatever be the humidity of the soil in which they are constructed, water never penetrates them; the walls are nicely

[1] *Cteniza.*  [2] *Ct. Sauvagesii.*  [3] *Ct. fodiens.*

[4] The French word is *féyure*, which I cannot find in the dictionaries, but it means, the circular frame of the mouth of the tube which receives the door.

covered with a tapestry of silk, having usually the lustre of satin, and almost always of a dazzling whiteness. He mentions only four species of the genus as at present known. One which was found in the Island of Naxos;[1] another in Jamaica;[2] a third in Montpellier;[3] and a fourth, that which is the subject of his Memoir, in Corsica; to which I may add a fifth species, found frequently by Mr. Bennett, in different parts of New South Wales.[4]

The habitations of the species in question are found in an argillaceous kind of red earth, in which they bore tubes about three inches in depth, and ten lines in width. The walls of these tubes are not left just as they are bored, but they are covered with a kind of mortar, sufficiently solid to be easily separated from the mass that surrounds it. If the tube is divided longitudinally, besides this rough cast, it appears to be covered with a coat of fine mortar, which is as smooth and regular as if a trowel had been passed over it; this coat is very thin, and soft to the touch; but before this adroit workman lays it, she covers the coarser earthy plaster-work with some coarse web, upon which she glues her silken tapestry.

All this shews that she was directed in her

---

[1] *Cteniza ariana.*    [2] *Ct. nidulans.*    [3] *Ct. cœmentaria.*
[4] *Wanderings in N. S. Wales, &c.* i. 328.

work by a Wise Master; but the door that closes her apartment is still more remarkable in its structure. If her well was always left open, she would be subject to the intrusion of guests that would not, at all times, be welcome or safe; Providence, therefore, has instructed her to fabricate a very secure trap-door, which closes the mouth of it. To judge of this door by its outward appearance, we should think it was formed of a mass of earth coarsely worked, and covered internally by a solid web; which would appear sufficiently wonderful for an animal that seems to have no special organ for constructing it: but if it is divided vertically, it will be found a much more complicated fabric than its outward aspect indicates, for it is formed of more than thirty alternate layers of earth and web, emboxed, as it were, in each other, like a set of weights for small scales.

If these layers of web are examined, it will be seen that they all terminate in the hinge, so that the greater the volume of the door, the more powerful is the hinge. The frame in which the tube terminates above, and to which the door is adapted, is thick, and its thickness arises from the number of layers of which it consists, and which seem to correspond with those of the door; hence, the formation of the door, the hinge, and the frame, seem to be a simultaneous operation; except that in fabricating the first, the animal has to knead the earth, as well as to spin the layers of web. By this admirable arrangement, these parts always correspond with each other, and the strength of the hinge, and the thickness of the frame, will always be proportioned to the weight of the door.

The more carefully we study the arrangement of these parts, the more perfect does the work appear. If we examine the circular margin of the door, we shall find that it slopes inwards, so that it is not a transverse section of a cylinder, but of a cone, and on the other side, that the frame slopes outwards, so that the door exactly applies to it. By this structure, when the door is closed, the tube is not distinguishable from the rest of the soil, and this appears to be the reason that the door is formed with earth. Besides, by this structure also, the animal can more readily open and shut the door; by its conical shape it is much lighter than it would have been if cylindrical, and so more easily opened, and by its external inequalities, and mixture of web, the spider can more easily lay hold of it with its claws. Whether she enters her tube, or goes out, the door will shut of itself. This was proved by experiment, for though resistance, more or less, was experienced when it was opened, when left to itself, it always fell down, and closed the aperture. The advantage of this structure to the spider is evident, for whether it darts out upon its prey, or retreats from an enemy, it is not delayed by having to shut its door.

The interior surface of this cover to its tube is not rough and uneven like its exterior, but perfectly smooth and even, like the walls of the tube, being covered with a coating of white silk, but much more firm, and resembling parchment, and remarkable for a series of minute orifices,[1] placed in the side opposed to the hinge, and arranged in a semicircle; there are about thirty of these orifices, the object of which, M. Audoin conjectures, is to enable the animal to hold her door down, in any case of emergency, against external force, by the insertion of her claws into some of them.

The principal instruments by which this little animal performs her various operations, are her mandibles or cheliceres, and her spinners. The former, besides the two rows of tubercles, between which, when unemployed, her claw, or sting, is folded, has at the apex, on their inner side a number of strong spines.[2] As no one has ever seen her at work upon her habitation, it cannot be known exactly how these organs, and

---

[1] PLATE XI. B. FIG. 2. *a.*

[2] *Observations sur le nid d'une Araignée lu à l'Acad. des Sc. le 21 Juin*, 1830, *par M. Victor Audoin;* and *Ann. de la Soc. Ent. de France.* ii. 69.

probably her anterior legs, are employed in her
various manipulations.

I have, in my collection, a tube or nest of the
Jamaica trap-door spider,[1] consisting merely of
the web, which is much larger than that just
described, being more than six inches long, and
three quarters of an inch in diameter in the
narrowest part, but near the mouth more than
an inch.   In this species the trap-door is semicircular, having a sloping margin; it is lined,
as well as the upper part of the tube, with a
strong close web, resembling parchment.   I can
detect in it no series of orifices, but I see here
and there little holes where the claws appear
to have been inserted.   This door is entirely
formed of layers of web, without any intermixture of earth.

Mr. Bennett, in his *Wanderings, &c.*[2] gives
some interesting particulars of the species discovered by him in New South Wales.   He
describes the tube, as about an inch in diameter
at the mouth, and the lid as formed of web
incorporated with earth, and exactly fitting the
mouth of the tube, in this resembling the *pioneer*.
He heard of a person who used to amuse himself
with feeding one of these insects: when its
meal was finished, it would re-enter its habitation, and pull down the lid with one of its

<hr>

[1] PLATE XI. B. FIG. 4.   .        [2] i. 328.

claws. He further observes, that to discover their habitations when the lid is down, from its being so accurately fitted to the aperture, was very difficult.

Though the particulars I have here stated, of the history and habits of these subterranean spiders, demonstrate, in every respect, as far as we know them, the adaptation of means to an end, far above the intelligence of the animal that exhibits them ; yet fully to appreciate the Wisdom, and Power, and Goodness, that fabricated her, and instigated her to exercise these various arts, and to employ her power of spinning webs, in building the structures necessary for her security, as well as for the capture of her prey, we ought to be witnesses to all her proceedings, which would probably instruct us more fully why she forms so deep a tube, and one so nicely covered with a peculiar tapestry from the mouth to the bottom. One of these ends, is, doubtless, to keep her tube dry.

2. Various are the modes of capturing their prey, exercised by the *second* Tribe of spiders, which have only *two* gills, some fabricating webs of various kinds for that purpose, and others lying in wait for them, and catching them by mere agility. The first of these are called *weavers,*[1] and the last, *hunters.*[2]

Some of the former construct silken tubes of

[1] *Araneïdæ textoriæ.*  [2] *A. venatoriæ.*

an irregular texture, open at both ends, in which they conceal themselves. Of this description is one, remarkable for having only six eyes,[1] which sits at the mouth of her tube, with her four anterior legs out of it, reposing by their extremity upon as many fine threads, which diverge from the mouth of the tube as from a centre, and probably contribute to form the toils, or are connected with them, which De Geer observed her to construct in front of her den,[2] and in which large flies are taken, which, by means of her stout mandibles, she soon kills, and then sucks their juices.[3]

Another species,[4] which spins a similar web with diverging threads, forming so many snares, is remarkable for the pertinacity with which it clings to its tube. The most effectual way to expel it, is to put in a live ant; scarcely has it entered, when the spider, in a violent agitation, uses its utmost efforts to frighten the intruder; if the ant disregards its menaces, it rushes out precipitately, and does not stop till it is two or three inches distant, when it halts to watch the motions of the ant, which, usually, when disengaged from the web, falls to the ground; upon this taking place, the former reenters its tube backwards. This species, though driven

---

[1] *Segestria senoculata.*          [2] vii. 261.
[3] Walck. *Araneïd. de France.* 195.          [4] *Segestria perfida.*

from its habitation by so small an insect, will fearlessly attack the largest flies, and it has been seen even to seize a very active wasp.[1]

The webs of the *retiary* or geometric spiders, which belong to another division of the weavers, are so well known that it is not necessary to give a very detailed account of their proceedings; but as Mr. Blackwall, in a very interesting Memoir in the *Zoological Journal*,[2] has added much to our previous knowledge on this head, especially with respect to the spiral circumvolutions that distinguish the webs of the tribe in question, I shall abstract, as briefly as I can, the main features of his account. Having formed the foundation of her net, and drawn the skeleton of it, by spinning a number of rays converging to the centre, she next proceeds, setting out from that point, to spin a spiral line of unadhesive web, like that of the rays, which it intersects, and to which she attaches it, and after numerous circumvolutions, finishes it at the circumference. This line, in conjunction with the rays, serves as a scaffolding for her to walk over, and it also keeps the rays properly stretched. Her next labour is to spin a spiral or labyrinthiform line from the circumference towards the centre, but which stops somewhat short of it; this line is the most important part

---

[1] Walck. *Araneïd. de France*, 202.  [2] v. 181.

of the snare. It consists of a fine thread, studded with minute viscid globules, like dew, which by their adhesive quality retain the insects that fly into the net. The snare being thus finished, the little geometrician selects some concealed spot in its vicinity, where she constructs a cell, in which she may hide herself, and watch for game; of the capture of which, she is informed by the vibrations of a line of communication between her cell and the centre of her snare.

The insects that frequent the waters require predaceous animals to keep them within due limits, as well as those that inhabit the earth, and the water-spider[1] is one of the most remarkable upon whom that office is devolved by her Creator. To this end her instinct instructs her to fabricate a kind of *diving-bell* in the bosom of that element. She usually selects still waters for this purpose. Her house is an oval cocoon, filled with air, and lined with silk, from which threads issue in every direction, and are fastened to the surrounding plants; in this cocoon, which is open below, she watches for her prey, and even appears to pass the winter, when she closes the opening. It is most commonly, yet not always, entirely under water; but its inhabitant has filled it with air for her

---

[1] *Argyroneta aquatica.*

respiration, which enables her to live in it. She conveys the air to it in the following manner: she usually swims upon her back, when her abdomen is enveloped in a bubble of air, and appears like a globe of quicksilver; with this she enters her cocoon, and displacing an equal mass of water, again ascends for a second lading, till she has sufficiently filled her house with it, so as to expel all the water. The males construct similar habitations, by the same manœuvres. How these little animals can envelope their abdomen with an air-bubble, and retain it till they enter their cells, is still one of Nature's mysteries that have not been explained. We cannot help, however, admiring and adoring the Wisdom, Power, and Goodness manifested in this singular provision, enabling an animal that breathes the atmospheric air, to fill her house with it under the water; and which has instructed her in a secret art, by which she can clothe part of her body with air, as with a garment, which she can put off when it answers her purpose. This is a kind of attraction and repulsion that mocks all our inquiries.

Amongst the spiders called the *hunters*, and the *vagrants*, some seize their prey like the lion or the tiger, with the aid of few or no toils, by jumping upon them, when they come within their reach. I have often observed a white or yellowish species of crab-spider[1]—a tribe so called because their motions resemble those of crabs—which lies in wait for her prey in the blossoms of umbelliferous and other white-blossomed plants, and can scarcely be distinguished from them, which when a fly or other insect alights upon the flower, darts upon it before she is perceived.

There is a very common black and white spider,[2] amongst the *vagrants*, which may always be seen in summer, on sunny rails, window-sills, &c. : when one of these spiders, which are always upon the watch, spies a fly or a gnat at a distance, he approaches softly, step by step, and seems to measure the interval that separates him from it with his eye; and, if he judges that he is within reach, first fixing a thread to the spot on which he is stationed, by means of his fore feet, which are much longer and larger than the others, he darts upon his victim with such rapidity, and so true an aim, that he seldom misses it. Whether his station is vertical or horizontal is of little consequence, he can leap equally well from either, and in all directions. He is prevented from falling, by the thread just mentioned, which acts

---

[1] Related probably to *Thomisus citreus*.
[2] *Salticus scenicus*.

as a kind of anchor, and enables him to recover his station, when without such a help he would be, as it were, driven out to sea.

We see in these latter instances, that though the art and means of weaving snares to entrap their prey have not been granted to these hunters and vagrants, yet that their Creator has endowed them with increase of agility, and the power of moving, without turning round, in all directions, which fully make up to them for that want.

Before I conclude this history of spiders, I must mention a very remarkable one, described and figured by Freycinet, under the name of *Aranea notacantha*,[1] but which appears to belong to no known genus of the Order. It is stated to have at its posterior extremity a long cylindrical tube, terminated by two eyes!! But this, surely, must be a mistake. At the anterior part of the thorax are four eyes, in a square, and one on each side. The form of the abdomen and its tube are very remarkable. This spider was found in a small island near Port Jackson, in an irregular web attached to the shrubs.

2. The *Pedipalps*, forming the *second* Order of Arachnidans, will not detain us long. The principal animals belonging to it, are the *scor-*

---

[1] PLATE XI. FIG. 2.

*pions*, which are not only remarkable for the
powerful organs by which they are enabled
to seize their prey, but also for their jointed
tail terminating in a deadly sting. Their aspect
alone, when they are moving with their open
forceps advanced before their head, and their
tail turned over it, is enough to create no
little alarm in the beholder; and if he were
told that one genus of the tribe goes by the
name of *man-killer*,[1] and should read in Aristotle,
that though some were harmless, the sting of
others was fatal both to man and beast,[2] the
degree of his alarm would not be diminished.
But though the venom of these creatures, when
provoked and put upon self-defence, may sometimes prove fatal to man and the higher animals,
yet this is not the main purpose for which their
Creator has given them such means of annoyance.
Their food consists of various beetles and other
insects, arachnidans, and wood-lice; many of
which they could not easily master and devour,
after they have seized them with their forceps,
without the aid of their tail and its sting;[3] this
they can turn over their head, and moving it
in any direction, immediately kill their prey,

---

[1] *Androctonus.*

[2] *Hist. Animal.* l. viii. c. 39. Comp. *N. D. D'Hist. Nat.*
xxx. 431.

[3] See above, p. 233.

however strong and active, by the fatal venom it instills.

Our Saviour alludes to the scorpion as one of the symbols of the evil spirit: and as a zodiacal sign with the Egyptians, it represented Typhon, which seems to prove that our Saviour's application of it was in conformity with a current opinion.

The other Pedipalps,[1] though one of them has a jointed tail like the scorpions,[2] are not armed with a sting. Probably the animals that they feed upon offer less resistance than the prey of the latter.

With regard to the Arachnidans in general, the object of their creation appears to have been to assist in keeping within due bounds the insect population of the globe. The members of this great and interesting Class are so given to multiply beyond all bounds, that were it not for the various animals that are directed by the law of their Creator to make them their food, the whole Creation, at least the organized members of it, would suffer great injury, if not total destruction, from the myriad forms that would invest the face of universal nature with a living veil of animal and plant devourers. To prevent this sad catastrophe, it was given in

---

[1] *Phrynus, &c.*       [2] *Thelyphonus.*

charge to the spiders, to set traps everywhere,
and to weave their pensile toils, from branch
to branch, and from tree to tree, and even to
dive under the waters.   And, more particularly,
to them we are mainly indebted for our de-
liverance from a plague of *flies* of every description, which, if the spiders were removed,
of which they form the principal food, would
subject us to incredible annoyance.[1]

The scorpions, and other Pedipalps, are
found only in warm climates, where they are
often very numerous, and, like the centipedes,
creep into beds.[1]   Insects multiply, beyond
conception, in such climates, and unless Providence had reinforced his army of insectivorous
animals, it would have been impossible to exist
in tropical regions.   The animals we are speaking of, not only destroy all kinds of beetles,
grass-hoppers, and other insects, but also their
larves, and even eggs.

### *Pseudarachnidan Condylopes.*

This Class, which is formed from the *Tracheary
Arachnidans* of Latreille, differs from the preceding principally in the organs of *Respiration*
and *Circulation.*

BODY coriaceous, or crustaceous.   *Spiracles*

[1] See above, p. 68.

connecting with *tracheæ* for respiration. *Circu-lation* obscure. *Eyes* 2—4. *Legs* 6—8. *Sexual organs* single.

The Class consists of two Orders, perfectly analogous to those of the Arachnidans, which may be denominated, *Pseudo-scorpions* and *Phalangidans.*

1. *Pseudo-scorpions.* BODY oblong, divided into several segments. *Eyes* 2—4. *Legs* 6—8.

2. *Phalangidans.* BODY consisting of one segment, with the analogue of the abdomen consisting of folds. *Eyes* 2. *Legs* 8, elongated.

1. I have already given an account of the most interesting genus of this Order, the *Sol-puga*, on a former occasion;[1] and there is little known of the history of the *book-crabs*,[2] except that they are often found in books; I have also occasionally met with them in the drawers of my insect cabinets, moving slowly on, with their arms expanded, probably they were in search of the mite that is so injurious to specimens of insects; they are also often found upon flies. One genus,[3] in this tribe, has *four* eyes, all the rest of the Class have only *two.*

2. The most remarkable genus[4] of the second Order of Pseudarachnidans is one described in the *Linnean Transactions*,[5] in which the posterior

---

[1] See above, p. 85.  [2] *Chelifer*  [3] *Obisium.*
[4] *Gonyleptes.* K.  [5] xii. 450. *t.* xxii. *f.* 16.

legs exhibit a raptorious character, and seem fitted either to seize or retain their prey. The common Phalangidans, or harvest-men, have been treated of in another place.[1]

The animals of this class seem to be universally insectivorous, though fabricating no snares.

### Acaridan Condylopes.

We are now arrived at a Class of Condylopes, that, with respect to their *food*, have a much more extensive commission than those which we have lately considered, the Arachnidans, and Pseudarachnidans. Under the name of *mites* they are universally known, and when some of our most essential articles of food, as cheese and flour, get old, or in any degree musty, they soon swarm with these minute animals, which, wherever they are established, multiply beyond conception; mites also attack not only decaying substances, but also living ones; in man they are the cause of a most revolting distemper;[2] under the name of ticks they attack dogs and other animals, and few insects altogether escape from their annoyance; and they not only infest the inhabitants of the earth and air, but are also found swimming in every pool; so that their field of action seems to be the whole creation of organized beings.

[1] See above, p. 90.  [2] See *The Lancet*, i. 1834-5. 59.

The Class may be thus characterized:

BODY without any insection or impression marking out its parts, consisting of a single segment, and without folds. *Mouth* and organs various. *Eyes* 2. *Legs* 6—8, short.

Latreille has divided this Class, including in it the preceding one, into *seven* Families; but perhaps it would be better to consider it as divided into *two* Orders, *mites*,[1] and *ticks*,[2] or those that do not *suck* their food, and those that are fitted with an organ adapted to suction.

I shall select an instance or two from animals of this Class, which shew the care of the Creator, for these little beings apparently so low in the scale of Creation; His foresight of every circumstance in which they would be placed; and His adaptation of their structure to their assigned station.

This is particularly conspicuous in the case of a species of bat-mite,[3] which was first noticed by one of our most celebrated microscopical observers, Mr. Baker, and has since fallen under the notice of M. V. Audoin, well known for his acute investigation of the external parts of Insects, who kindly sent me a memoir of his on this and other Acaridans, extracted from the *Annales des Sciences Naturelles* for the year 1832. If we consider the animal that this mite

---

[1] *Acari.*      [2] *Ricini.*      [3] *Pteroptes.*

inhabits, the bat, and that it affords much less shelter than the birds, to any parasite that may be attached to it, especially as the species that I am speaking of is stated usually to fix itself to the membrane of the wings, which being a naked membrane, would seem to expose it to be easily shaken off when the animal is flying: we easily comprehend that it stands in need of some particular provision to counteract this circumstance.

Like those of many other mites, its feet are furnished with a vesicle which is capable of contraction and dilatation, and which the animal can probably use as a sucker to fix itself; but if by any sudden jerk it is unfixed, to prevent its falling, it is gifted with the power of turning upwards, in an instant, two, four, six, or even all its legs, according to circumstances, sufficiently to support itself, and can walk in this position, as it were upon its back, as well as it does in the ordinary way with that part upwards; it may be often seen with four turned upwards while it walks upon the other four,[1] so that it is ready, upon any accident, instantaneously to use them, and to lay hold of the wing.

The bat is infested by another parasite, placed by Dr. Leach at the end of the *Acaridans*, and by Latreille, but not without hesitation, after the

---

[1] Baker *on Micr.* ii. 407. *t.* xv. *f.* E. F. G.

*Diptera.* I may therefore be justified in introducing the animal in question here, since, inhabiting the same subject, their proceedings will serve to illustrate each other, and to demonstrate the agency and design of the Supreme Cause in the concurring structure of these parasites. The one I here allude to may be called the *bat-louse.*[1] Latreille, who has described very minutely a species of this genus,[2] informs us that their head is implanted in a singular situation, the back of the thorax, between the middle and the anterior extremity,[3] immediately behind the part to which the anterior legs are attached. The middle of the back, in the common species, presents a cavity, which terminates posteriorly in a kind of pouch,[4] so that the head can be thrown back and its extremity received by it. From this situation, it is evident that the animal cannot take its nutriment from the bat in the ordinary position, with the back upwards; it must, therefore, necessarily stand with it downwards when engaged in suction. When under the forming hand of the Almighty Creator, its legs were planted, it was not on the *lower* side of the trunk, as they usually are in other hexapods, but on the *upper* side or margin of that part.[5] Colonel Montague observes,—" So strange and

---

[1] *Nycteribia,* Lat.　　　　[2] *N. Blainvillii.*
[3] See Montague. *Linn. Trans.* xi. *t.* iii. *f.* 5.
[4] *N. Verpertilionis.*　　[5] *N. D. D'Hist. Nat.* xxxiii. 131, 132.

contradictory to experience is the formation of
this Insect, that were it not for the structure
of the legs, no one could doubt that the upper
was actually the under part of the body.[1]  From
the account given by the last acute and indefatigable naturalist, the motions of this little
creature are so rapid as to be almost like flight,
and it can fix itself in an instant wherever
it pleases.  Putting some into a phial, their
agility was inconceivable; not being able, like
other Dipterous insects, to walk upon the glass,
their efforts were confined to laying hold of each
other, and during the struggle they appeared
flying in circles."[2]

Their head is furnished with antennæ and
feelers, immediately below the insertion of the
former, on each side, is a slightly prominent
eye, so that they have sight to guide them in
their motions, which the *bat-mite* appears to be
without.

I may conclude this account with the pious
reflection of the worthy author lately mentioned.
The very singular structure of this insect,
which, at first, appears to be a strange deformity
in nature, and excites our astonishment, will,
like all other creatures, constructed by the same
Omnipotent hand, be found to be most admirably contrived for all the purposes of its

[1] *Linn. Tr.* xi. 12.        [2] Ibid. 13.

creation; and the scrutinizing naturalist will soon discover this unusual conformation to be the character which at once stamps its habits and economy.[1]

One of the most singular animals of this Class is one called the *vegetating* mite.[2] These are fixed for a time, by an anal thread, to certain beetles, by means of which, as by an umbilical chord, they derive their nutriment from them. After a certain time, they disengage themselves, and seek their food in the common way of their tribe.

It is difficult to say where Latreille's Order of *Aporobranchians*[3] should properly be placed. Savigny considers them as leading from the Crustaceans to the Arachnidans by *Phalangium*. If they are parasitic upon marine animals, as there is reason to believe, might they not, in some sort, be regarded as one of those branches, which, without going by the regular road, form a link between tribes apparently distant from each other?[4] They seem, in some respects at least, to present an analogy, if not an affinity, to the Hexapod parasites, the bird-louse,[5] &c. I offer this merely as a conjecture.

---

[1] *Linn. Tr.* xi. 13.     [2] *Uropoda vegetans.*
[3] *Nymphon. Pycnogonum,* &c.     [4] See above, p. 18.
[5] *Nirmus.*

## Chapter XX.

*Functions and Instincts.    Insect Condylopes.*

The animals of the class we are next to consider,
have been regarded by many modern zoologists,
especially of the French school, as inferior both
to Crustaceans and Arachnidans, on account of
their having only, as it were, a rudimental heart,
exhibiting indeed a kind of systole and diastole,
but unaccompanied by any system of vessels by
which the blood might circulate in them. A
learned and acute writer, and eminent zoologist,
amongst our own countrymen, has with great
force controverted the justice of this sentence of
degradation pronounced upon Insects; an opinion which has also been embraced by many
other modern writers on the subject, and considerable doubt has been shown to rest upon the
main foundations upon which the illustrious and
lamented Baron Cuvier, who was the father of
the hypothesis, had built it.[1]

But the important discoveries of Dr. Carus,
who first proved that a *circulation* really exists
in various larves of Insects, and afterwards that

[1] Mac Leay, *Hor. Entomolog.* 204, 297.

it is also discoverable in several perfect ones,[1] have placed the matter beyond all doubt. Taking, therefore, into consideration the *nervous* system of Insects, as well as those of circulation and respiration, as ought, in all reason, to be done —for upon comparison of these three systems so intimately connected with life and sensation, surely the first place is due to that by which alone the animal is conscious of its existence and that of the world it inhabits, and is enabled to run the race appointed by its Creator; surely if even no Carus had appeared to demonstrate the existence of a circulation in these animals, still the perfection of their nervous system, compared with that of the Molluscans, in determining their respective stations, would be a sufficient counterpoise to a heart and vascular system for circulation; and if to this superiority we add the number and nature of the several organs by which this system acts, and the fruits of such agency in the activity and various instincts of the animals endowed with it, embodying the moving will, the informing sense, the impelling appetite, compared with the inertness and sluggish motions, and apathetic existence, and paucity of instinctive actions in the great majority of the Molluscans,—who is there that will hesitate to conclude that He who created the *Insect*

[1] *Introd. to Comp. Anat. E. T. by Gore*, ii. 392. *Act. Acad. Cæs Nat. Cur.* xv. ii.

world, gifted them with so many and such wonderful instincts, inspired them with such inces-
sant activity, fitted them with such various
organs for such a diversity of locomotions under
the earth, on the earth, in the air and in the
water, meant to place them far above the headless *Oyster*, with scarcely any organs of sensation, and scarcely any motion but that of opening and shutting its shell, or even than the
*Cuttle-fish*, though furnished with eyes, and
even three hearts, and a very extraordinary
animal, yet destitute of many organs of the
senses and of locomotion found in Insects, and
most of those that they have not formed upon
the plan of the higher animals, but rather borrowed from the confessedly lower Classes of
Polypes and Radiaries?[1]

With regard to the *Crustaceans* and *Arachnidans*, setting aside the superiority of *Insects* in
their instincts, the single circumstance of the
*reproduction* of mutilated organs in the former
seems to prove an inferiority of rank and a tendency towards the Polype.[2]

When we consider attentively these little
beings, the infinite variety of their forms, the
multiplicity and diversity of their organs, whether of sense or motion, of offence or defence,
for mastication or suction ; or those constructed

[1] Vol. I. p. 303, 304.     [2] Mac Leay, *Hor. Ent.* 206, 298.

with a view to their several instincts, and the
exercise of those functions devolved upon them
by the wisdom of their Creator; the different
kinds also of sculpture which is the distinction
of one tribe, and of painting, which ornaments
another, the brilliant colours, the metallic lustre,
the shining gold and silver with which a liberal
and powerful hand has invested or bespangled
numbers of them; the down, the hair, the wool,
the scales, with which He, who careth for the
smallest and seemingly most insignificant works
of his hand, hath clothed and covered them;
when all these things strike upon our senses,
and become the subject of our thoughts and
reflection, we find a scene passing before us far
exceeding any, or all of those, that we have
hitherto contemplated in our progress from the
lowest towards the highest members of the animal kingdom, and which for its extent, and the
myriads of its mustered armies, each corps distinguished as it were by its own banner, and
under its proper leaders, infinitely outnumbers
all the members of the higher Classes, which
stand as it were between aquatic and terrestrial
animals, many of its tribes under one form inhabiting the water, and under another the earth
and the air.

The following characters distinguish this great
Class :

BODY, covered with a horny or coriaceous integument. *Spinal chord* knotty, terminating anteriorly in a bilobed brain; a *heart* and imperfect *circulation*, sometimes vascular, and sometimes extra-vascular; *respiration* by *tracheæ*, receiving the air by *spiracles; legs* jointed, in the perfect insect always *six*.

The Class of Insects may be divided into two *Sub-classes,*[1] viz. *Ametabolians*, or those that do *not* undergo any *metamorphosis*, and have no wings; and *Metabolians*, or those that undergo a *metamorphosis*, and are usually fitted with wings in their final state.

*Sub-class* 1.—*Ametabolians* are further subdivided into two *Orders, Thysanurans* and *Parasites.*

*Order* 1.—The *Thysanurans* are remarkable for their anal appendages, which consist either of jointed organs resembling antennæ, and approaching very near to the caudal organs of the cockroach,[2] the use of which is not certainly known; or of an inflexed elastic caudal fork bent under the abdomen, which enables them to leap with great agility. To the first of these tribes belongs the common *sugar-louse,*[3] and to the last the *spring-tails.*[4]

It must be observed, however, that this is not

---

[1] See above, p. 18.    [2] *Blatta.*
[3] *Lepisma.*    [4] *Podura. Sminthurus.*

a *natural* Order, for there is no analogy between the jointed tails of the sugar-louse, which some have supposed to belong or approach to the *Orthoptera*, and the unjointed leaping organ of the spring-tail. The latter animals, indeed, seem to form an *osculant* tribe, without the pale of the Class of Insects, and perhaps having some reference to the *Chilopodans* amongst the Myriapods, with which they agree, in having only *simple* eyes, like spiders, on each side of the head. Those of the spring-tails consist of eight such eyes, arranged in a double series, and planted in an oval space, in shape resembling an Insect's eye. The Chilopodans have only four on each side. The Insects of this Order probably feed upon detritus, whether animal or vegetable, their masticating organs being very weak, and fitted to comminute only putrescent substances.

*Order* 2.—The Order of *Parasites*—consisting of the most unclean and disgusting animals of the whole Class, infest both man, beast, and bird, and no less than four[1] species, accounted by Linné, &c. as varieties, being attached to the former—may be divided into two sections, those that live by suction, and those that masticate their food. To the first of these belong the

[1] *Pediculus Capitis, Corporis, Nigritarum*, and *Phthirus Pubis.*

human and the dog-louse, and to the other the various lice that inhabit the birds,[1] of which almost every species has a peculiar one.

I have, on a former occasion, alluded to the Order of *Parasites*, when speaking of punitive animals:[2] here I must observe, that like other instruments employed by God to visit the sins of mankind, they are intended to produce a *sanative* effect, as well as to punish.[3] It is generally known that they abound only on those whose habits are dirty, in whom they may prevent the diseases which such habits would otherwise generate, as well as stimulate them to greater attention to personal cleanliness. The *bird-louse* is probably useful to birds in devouring the sordes which must accumulate at the root of their plumes.

*Sub-class* 2.—*Metabolians*, by most modern writers on Insects, are considered, from their oral organs, as constituting *two* Sections, which are denominated *Haustellate* and *Mandibulate* Insects. I may here observe that the instrument of suction in a *Haustellate* mouth consists of pieces, though differently circumstanced, precisely analogous to those employed in mastication in a *Mandibulate* one, which has been most

---

[1] *Nirmus.*
[2] Vol. I. p. 12.   See *Introd. to Ent.* i. 83.
[3] Ibid, p. 253.

satisfactorily demonstrated, and with great elegance, by M. Savigny, in the first part of his *Animaux sans Vertèbres*.[1]

As there are several Orders called *Osculant*, that are intermediate between these Sections, I shall arrange the whole in three columns.

OSCULANT ORDERS.

1. *Aphaniptera.*
2. *Homaloptera.*
3. *Trichoptera.*
4. *Dermaptera.*
5. *Strepsiptera.*

| HAUSTELLATE ORDERS. | MANDIBULATE ORDERS. |
|---|---|
| 6. *Diptera.* | 10. *Hymenoptera.* |
| 7. *Lepidoptera.* | 11. *Neuroptera.* |
| 8. *Homoptera.* | 12. *Orthoptera.* |
| 9. *Hemiptera.* | 13. *Coleoptera.* |

With regard to the characters of these Orders :

*Order* 1.—-The *Aphaniptera* (*Flea, Chigoe*) are apterous and parasitic, but differ from the Order of *Parasites* by undergoing a metamorphosis. They connect the *Suctorious Parasites* with the *Diptera.*

*Order* 2.—The *Homaloptera* (*Forest-fly, &c.*) called also *Pupipara,* because their eggs are hatched in the matrix of the mother, where they pass their larve state, and are not excluded till

[1] t. i.—iv.

they have become pupes. Most of them have two wings, but one genus is apterous:[1] these seem intermediate between certain *Acaridans*, as the bat-mite, and the *Diptera;* they seem also, in some respects, to connect with the *Arachnidans*, whence they have been called *spider-flies*.

*Order* 3.—The *Trichoptera* (*Caseworm-flies*) have four *hairy* membranous wings, in their nervures resembling those of *Lepidoptera*, the under ones folding longitudinally. The mouth has four palpi, but the masticating organs are merely rudimental. Their place seems to be somewhere between the *saw-flies* and those *moths* whose caterpillars clothe themselves with different substances.

*Order* 4.—The *Dermaptera* (*Earwigs*) have two elytra and two wings of membrane, folded longitudinally, and their tail is armed with a forceps. They appear to be between the *Coleoptera* and *Orthoptera*.

*Order* 5. — The *Strepsiptera* (*Wild bee-fly*, *Wasp-fly*), parasitic animals, that have two ample wings, forming the quadrant of a circle, and of a substance between coriaceous and membranous, and two elytriform subspiral organs, appendages of the base of the anterior legs. Their place is uncertain, some placing them

---

[1] *Melophagus.* The *Sheep-louse.*

between the *Coleoptera* and *Dermaptera;* and others between the *Lepidoptera* and *Diptera.*

---

*Order* 6.—The *Diptera,* (*Two-winged Flies and Gnats,* &c.) as their name indicates, have only *two* membranous wings, usually accompanied by *two winglets,* representing the under wings of the Tetrapterous Orders, and *two poisers,* which appear connected with a spiracle.

*Order* 7.—The *Lepidoptera* (*Butterflies and Moths*) have *four* membranous wings, covered with minute scales, varying in shape.

*Order* 8.—The *Homoptera* (*Tree-Locusts, Frog-hoppers, Froth-hoppers*) have four deflexed wings, often of a substance between coriaceous and membranous.

*Order* 9.—The *Hemiptera* (*Bugs,* &c.) have four organs of flight, the upper pair being horny or coriaceous, but tipped, in the generality, with membrane, the lower pair being membranous.

---

*Order* 10.—The *Hymenoptera,* (*Saw Flies, Gall Flies, Ichneumon Flies, Bees, Wasps, Ants,* &c.) which are the analogues of the *Diptera,* have four membranous wings, and the tail of the female is usually armed with a sting, or instrument useful in laying their eggs.

*Order* 11.— The *Neuroptera* (*Dragon Flies, Lace-winged Flies, Ephemeral Flies, White Ants,* &c.) have four membranous wings, usually reticulated by numerous nervures, but no sting or ovipositor. They are analogues, especially *Ascalaphus*, of the *Lepidoptera.*

*Order* 12.—The *Orthoptera* (*Cockroaches, Locusts, Praying-insects, Spectres, Grasshoppers, Crickets,* &c.) have mostly two *tegmina,* or upper wings, of a substance between coriaceous and membranous, and two under ones, formed of membrane, and folded longitudinally when unemployed. These are analogues of the *Homoptera.*

*Order* 13.—The *Coleoptera (Beetles)* have two upper organs, of a horny or leathery substance, called *elytra,* to cover their two membranous wings, which are folded longitudinally and transversely. These are analogues of the *Hemiptera,* especially those with no apical membrane.

———

In considering the three descriptions of Orders here enumerated and characterized, it must be recollected that we are not following the usual order of arrangement in systems, that of *descending* from the highest to the lowest; but that we are *ascending* in an inverse direction, consequently that, in the above tables, the *lowest* numbers indicate the *lowest* and not the *highest* Orders.

I shall now make some remarks, as to their *functions* and *uses,* upon the animals constituting these several Orders, enlivening them

occasionally with such histories, not before produced, or not well known, as may interest the
reader and answer the great end of this treatise,
the glory of God, as manifested in the history
and instincts of animals.

Before, however, I enter upon the separate
consideration of these Orders, I must premise
a few remarks upon the circumstance which distinguishes them from the preceding Sub-class,
their *metamorphoses.* I have, on a former occasion,[1] mentioned some beneficial effects resulting from this law of the Creator; and its
action and the results of it have been so ably
explained and illustrated in another treatise,[2]
that it is quite unnecessary for me to enter
largely into the subject. The striking remarks
made upon the developements of the higher
animals, towards the close of the treatise alluded
to,[3] merit particular attention.

It has been observed by an ingenious and
learned writer[4] on this subject—that every species of plant, in the course of the year, exhibits itself in different states. First are seen the succulent stems adorned with the young foliage, next
emerge the buds of the flowers, then the calyx
opens, and permits the tender and lovely blossoms to expand. The insects destined to feed
upon each plant must be simultaneous in their

---

[1] See above, p. 25.    [2] Roget. *B. T.* i. 302—316.
[3] *Ibid.* ii. 631.    [4] Dr. Virey.

developement. If the butterfly came forth before there were any flowers, she would in vain
search for the nectar that forms her food; and if
the caterpillar was hatched after the leaves had
begun to fade and wither, she could not exercise
her function.[1] In another passage he thus illustrates this analogy between the metamorphoses
of the insect and the successive developements of
the plant. If we first place an egg, says he,
next to it its caterpillar, further on its chrysalis,
and lastly the butterfly; what have we but an
animal stem, an elongation perfectly analogous
to that of the plant proceeding from its seed, by
its stem and its appendages to the bud, the
blossom, and the seed again?[2] For the different
kinds and forms of larves and pupes I must
refer the reader to another work,[3] merely observing that, in their forms, the larves seem to
represent all the preceding Classes of Condylopes, and also some Annelidans and Molluscans.
The great majority of pupes are not locomotive,
and take no food, while the rest are locomotive
and continue to feed. This circumstance sometimes exposes the former to the attacks of their
enemies, the ichneumons, and thus numbers are
destroyed which would otherwise escape; but
though, in this state, they are thus more exposed
to the attack of one enemy, they are more ef-

[1] *N. D. D'H. N.* xx. 348.        [2] *Ibid.* 355.
[3] *Introd. to Ent.* iii. Lett. xxx. xxxi.

fectually concealed from those of another, the insectivorous birds. Those that bury themselves in the earth seem still more privileged from attack.

*Orders* 1, 2, and 6. There is so close a connection between the *fleas*, the *pupiparous insects*, and the *two-winged flies*, that it will be best to consider them under one head. The former of these, the fleas,[1] the mosquitos, or gnats,[2] and the horse-flies,[3] all suck the blood of man, as well as that of beast or bird.[4] The wonderful strength and agility of the *flea* are well known,[5] and it appears to have been endowed with those faculties by its Creator, to render its change of station from one animal to another, and means of escape, more easy; and though the bite of mosquitos, and other blood-suckers, is, at certain times of the year and in certain climates, an almost intolerable annoyance;[6] yet, doubtless, some good end is answered by it; with regard to *cattle*, it is evident that, while they are suffering from the attack of these blood-letters, their feeding is more or less interrupted; a circumstance which may be attended by beneficial effects to their health; and probably even to man, the torment he experiences may be compensated, in

---

[1] *Pulex.*    [2] *Culex.*    [3] *Tabanus. Stomoxys.*
[4] *Introd. to Ent.* 1. 100, 109, 112, &c.
[5] *Ibid.* ii. 310. iv. 195.    [6] *Ibid.* 113.

a way that he is not aware of, on account of which, principally, a wise Physician prescribed the painful operation, and furnished his chirurgical operators with the necessary and indeed most curious knives and lancets.

Another group connecting the *bat-mite* and *bat-louse*, and the *Arachnidans*, perhaps, with the *Diptera*, are those two-winged insects, called *pupiparous* or *nymphiparous*, because their young when extruded from the abdomen of the mother, though appearing like eggs, are really in the state of *nymph* or *pupe*. It is remarked of this group, which is parasitic upon beasts and birds, that its internal structure is particularly accommodated to this circumstance; it is furnished with a regular matrix, consisting of a large musculo-membranous pocket, and with ovaries totally different from those of other insects; but, by their configuration and position, exhibiting a considerable resemblance to those of a woman.[1] The reason of this singular aberration from the gestation of other *Diptera*, which, with few exceptions, are oviparous, seems connected with their peculiar habits: in their perfect state they are usually winged, and attach themselves externally to horses, oxen, &c.; it may therefore be the means of preserving the race from extinction, that they are supported in the womb of

---

[1] Latr. *Crust. Arachn. et Ins.* ii. 542.

their mother, in some inscrutable way, during their grub state, and only leave her when their next change will enable them readily to attach themselves to their destined food.

The *gad-flies*,[1] though they do not, like the forest flies, nourish their young in their own womb; yet their Creator instructs some of them to deposit their eggs in a situation where means are provided for their conveyance to a more capacious matrix, ministering to them a copious supply of lymph, which forms their nutriment, in the stomach and intestines of the horse, for this animal, with its own mouth, licks off the eggs, wisely attached, by this fly, to the hairs of its legs in such parts as are exposed to this action; and thus unwittingly, itself conducts its foes into its citadel: others of the same genus undermine the skin of the ox, of the sheep, and in some countries, even of man himself. The grubs, by their action in their several stations, produce a purulent matter, which they imbibe, and which is stated by those who have studied them, to be beneficial to the animals they attack.[2] Another tribe of this Order, the *flesh-flies*,[3] lay their eggs on dead bodies, and soon remove those nuisances, and the putrid and

---

[1] *Œstrus, &c.*

[2] The species of gad-flies here alluded to are *Gastrus Equi*, and *Œstrus Bovis, Œ. Ovis* and *Œ. Hominis.*

[3] *Sarchophaga.*

pestilential miasmata which they occasion, from the face of our globe. This function is of such importance to the welfare of our species, that some of these *flies*, in order that no time may be lost, are viviparous,[1] and bring forth their young in a state in which they can begin their work as soon as they are born.

The *aphidivorous flies*[2] have another function, in conjunction with the *lace-winged flies*,[3] *lady-birds*,[4] and some other insects, to reduce and keep within due limits the infinite myriads of the *plant-lice*,[5] which, in these climates, are the universal pests of the garden, the orchard, and the field. There are also flies[6] that lay their eggs in the combs of *humble-bees*, which, as it were, wear their livery, for the hairs that clothe their body are so disposed and coloured, as to imitate that of the bee, whose nests they frequent; so that, probably, they are often mistaken for members of the family, and effect their mischief unmolested.

Another tribe of flies, called *hornet-flies*,[7] with some others related to them,[8] like a hawk or other predaceous bird, seize their prey with their legs, or their beak,[9] but it can only be with

---

[1] *Se-vivipara.*
[2] *Syrphus, &c.*
[3] *Hemerobius.*
[4] *Coccinella.*
[5] *Aphides.*
[6] *Volucella, &c.*
[7] *Asilus.*
[8] *Empis.*
[9] *Introd. to Ent.* i. 274.

the view of sucking its juices, as they have no masticating organs.

Dipterous insects, however, are not confined to *animal* food, whether living or putrescent, many also subsist upon a *vegetable* diet. Mushrooms and other agarics sometimes swarm with the grubs of certain flies or gnats;[1] others pass their first states in decaying timber; the narcissus and onion flies[2] feast upon the bulbs from which they take their name; and a little gnat,[3] when a grub, feeds upon the pollen of the flowers of the wheat.

To these may be added those flies, that in their first state, may be regarded as purifiers of stagnant waters, and other offensive fluids or semi-fluids. The larves of the *gnat* or mosquito are aquatic animals which may be seen either suspended at the surface, or sinking in most stagnant waters, compensate in some degree, for the torment of their blood-thirsty attacks, by discharging this function, and assisting to cleanse our stagnant waters from principles that might otherwise generate infection. A variety of others contribute their efforts to bring about the same beneficial purpose. Almost all the *Diptera*, in their perfect state, even the blood-suckers, emulate the bees, in imbibing the nectar

---

[1] *Mycetophila, &c.*

[2] *Eristalis Narcissi*, and *Scatophaya Ceparum*.

[3] *Cecidomyia Tritici.*

from the various flowers with which God has
decorated the earth, and thus assist in keeping
within due limits, the, otherwise suffocating,
sweets that they exhale.

From the statement here given, we see that
the Creator has provided the members of this
Order with a very diversified bill of fare, and
that their efforts in their several states, and
various departments, are of the first importance,
as scavengers and depurators, to remove or miti-
gate nuisances, that would otherwise deform and
tend to depopulate our globe.   What they want
in volume, is compensated for by numbers, for
perhaps the *individuals* of no Order are so numerous.   It is true, in particular periods, the locusts
and aphides seem to outnumber them ; yet, ordinarily, the two-winged race, are those which
everywhere most force themselves upon our
attention ; during nearly three-fourths of the
year we hear their hum, and see their motions,
in our apartments, and even in the depth of
winter, in sunny weather, by their myriads,
dancing up and down under every hedge, they
catch our attention in our walks.

*Order* 10.—If we next turn our attention to
the *mandibulate* Order, which stands most in
contrast with the *Diptera*, the *Hymenoptera*
immediately occurs to us, in which we find a
variety of forms, which seem made to imitate

those of flies, or vice versa. Thus there are flies[1] that resemble saw-flies; others that simulate the ichneumonidan parasites; others again that resemble wasps, bees, and humble-bees.

Though the Insects belonging to this Order are included in the mandibulate Section; for their mouth is furnished with mandibles and maxillæ; yet they do not generally use them to *masticate* their *food*, but for purposes usually connected with their sequence of instincts, as the bees in building their cells;[2] the wasps in scraping particles of wood from posts and rails for a similar purpose, and likewise to seize their prey; but the great instrument by which, in their perfect state, they collect their food is their *tongue*, this, the bees particularly have the power of inflating, and can wipe with it both concave and convex surfaces; and with it they, as it were, *lick*, but not *suck*, the honey from the blossoms, for, as Reaumur has proved, this organ acts as a *tongue* and not as a *pump*.[3] In the numerous tribes that compose this most interesting of the Orders, the tongue is lambent, and varies considerably in its structure, but in the great majority it is a flattish organ, often divided into several lobes.

Some entomological writers have bestowed upon the members of the present Order the title

---

[1] *Aspistes*, Meig.   [2] See above, p. 188.   [3] *Mem.* &c. v. 322.

of *Principes*, as if they were the *princes* of the
Class of Insects, and if we consider the conspicuous manifestation of the Divine attributes
of Power, Wisdom, and Goodness, exhibited in
the wonderful instincts of those of them that are
*gregarious*, we shall readily concede to them this
title. If superior wisdom and devotedness to
the general good are the best titles to rank and
station ; the laborious and indefatigable *ant*, and
the *bee*, celebrated from the earliest ages for its
wonderful economy, its admirable structures,
and its useful products, are surely entitled to it,
though they cannot vie with the insects of many
of the other Orders in size, and in the brilliancy
and variety of their colours, and the pencil of
the Creator has not decorated their wings with
the diversified paintings which adorn those of
the butterfly.

The functions which are given in charge to
the several members of this Order are various.
Some, like the predaceous and carnivorous tribes
of the *Diptera*, appear engaged in perpetual
warfare with other insects ; thus the *wasps* and
*hornets* seize flies of every kind that come in
their way, and will even attack the meat in the
shambles; the *caterpillar-wasp*[1] walks off with
*caterpillars*, the *spider-wasp*[2] with *spiders*, and the
*fly-wasp*[3] with *flies*.   But the motive that influ-

---

[1] *Ammophila.*        [2] *Pompilus.*        [3] *Bembex.*

ences them, will furnish an excuse for their predatory habits. They do not commit these acts of violence to gratify their own thirst for blood, like many of the flies, but to furnish their young with food suited to their natures. The wasp carries the pieces of flesh she steals from the butcher to the young grubs in the cells of her paper mansion. The other wasps I mentioned each commit their eggs to the animal they are taught to select, and then bury it; so that the young grub when hatched may revel in plenty.[1]

Some of the *Hymenoptera* prefer a vegetable diet, and assist the *Lepidoptera* in their office. The caterpillars which infest many species of willow are hatched from the eggs of the *saw-flies;*[2] one genus[3] nearly related to them confines itself to timber, to which it is sometimes very destructive.

Another tribe affect plants in a very remarkable manner. Their egg-placer, like a magician's wand, is gifted with the privilege, by a slight puncture in the twig or leaf of any shrub or tree, or the stalk of any plant, to cause the production of a wonderful and monstrous excrescence, sometimes resembling moss, as in the Bedeguar of the rose, at other times, a kind of apple, or a transparent berry, both of which

[1] See *Introd. to Ent.* i. 346.
[2] *Cimbex, Tenthredo, Lyda, &c.* See *Introd. to Ent.* i. 255.
[3] *Sirex.*

seeming fruits, the oak, when touched by two of these little gall-flies of different species, produces as well as acorns: various other forms[1] their galls assume, which need not be here mentioned. It is to be observed that the eggs of these gall-flies grow after they are laid, and perhaps these singular productions are more favourable to their growth, being softer and more spongy and succulent than the twigs themselves would be. Even here Creative Power, Wisdom, and Goodness are conspicuously manifested, in providing such wonderful nests for these little germe-like eggs; these excrescences, indeed, instead of deforming the plants they are produced from, are often ornamental to them; and besides this are also, some of them, of the highest utility to mankind—witness the Aleppo oak-gall,[2] to which learning, commerce, the arts, and every individual who has a distant friend, are so deeply indebted.

Another tribe is equally useful in a different department; I allude to those Hymenoptera that are parasitic upon other Insects, particularly upon the destructive hordes of caterpillars that are often so injurious both to the horticulturist and agriculturist. These insects are denominated by Latreille *Pupivorous*, not, as some may suppose, because they devour

---

[1] *Introd. to Ent.* i. 446.    [2] *Cynips Scriptorum.*

insects in their second, or *pupe* state, but from
the *classical* meaning of the word, because they
devour them before they are arrived at their
perfect or adult state. This tribe may be considered as divided into two great bodies, one
represented by the *proper* Ichneumons of Linné,
which have, usually, veined wings, and the abdomen connected with the trunk by a footstalk ;
the other forming the *Minute Ichneumons* of that
great reviver of Natural History, distinguished,
usually, by having wings with few or no veins
on their disk, and by a sessile abdomen. These
attack eggs and chrysalises, as well as caterpillars. Though the latter are the principal,
yet they are not the only object of the great
Ichneumonidan host, for they attack insects of
every order indiscriminately ; they seem, however, to annoy beetles, grasshoppers, bugs, and
froghoppers, less than others. They may, with
great propriety, be called *conservatives*, since
they keep those under that would otherwise
destroy us.[1] A little fly, before alluded to in
these pages,[2] which appears very *destructive* to
*wheat* when in the ear, is rendered harmless, by
the goodness of Providence, by not less than
three of these little benefactors of our race.[3]

Connected with the subject of parasites is a
singular history communicated to me by the

---

[1] *Introd. to Ent.* i. 267.    [2] See above, p. 327.
[3] *Linn. Trans.* v. 107.

Rev. F. W. Hope, one of the most eminent entomologists of the present day. In the month of August, 1824, in the nest of a species of *wasp*,[1] he found more than fifty specimens of a singular little *beetle*, which may be called the *wasp-beetle*,[2] long known to frequent wasps' nests. From their being found in cells which were closed by a kind of operculum, he conjectures that they lay their eggs in the grub of the wasp, upon which they doubtless feed. Subsequent to this, upon opening some of the cells, he was surprised to find, instead of the beetles, several specimens of an Ichneumon belonging to Jurine's genus, *Anomalon*.[3] Upon another examination, some days after this, no more of these last insects appearing, he discovered that they had been pierced, in their chrysalis state, by a minute species belonging to the family of *Chalcididans*, of which he found no less than twenty specimens flying about in search of their prey.

" From the above facts," Mr. Hope remarks, " we have a convincing proof, if such were wanted, of a Superintending Power which ordains checks and counterchecks to remedy the superfecundity of the insect world." First the wasp, a great destroyer of flies and various other insects, and often a troublesome pest and an-

---

[1] *Vespa rufa.*        [2] *Ripiphorus paradoxus.*

[3] Latreille is of opinion that this is not a natural genus.  *N. D. D'H. N.* ii. 128.

noyance to man himself, is prevented from becoming too numerous, amongst other means, by the wasp-beetle; then, lest it should reduce their numbers so as to interfere with their efficiency, this last is kept in check by the *Anomalon*, which, in its turn, that it may obey the law, *Thus far shalt thou come, and no further*, becomes the prey of another devourer. Mr. Hope observed, and the fact is curious, that the specimens of the wasp-beetle obtained from the *female* wasps were about one-third larger than the others.

But of all the Hymenopterous, or indeed any other Insects, there are none, as I before observed, that illustrate the primary attributes of the Deity more strikingly than those that are *gregarious*, which build for the members of their societies spacious colleges, if I may so call them, capable often of containing many thousand inhabitants, and remarkable for the pains they bestow upon the nurture and education of their young. There are three great tribes in the present Order, distinguished by this instinct,—the *wasps* and *hornets*, the *bees* and *humble-bees*, and the *ants*.

The *wasps* and *hornets* are remarkable for the curious papier-maché edifices, in the construction of which they employ filaments of wood,—scraped from posts and rails with their own jaws,—mixed with saliva, of which the hexagonal cells, in which they rear their young, are formed, and often their combs are separated and supported by pillars of the same material; and the external walls of their nests are formed by foliaceous layers of their ligneous paper.[1] Latreille mentions a Brazilian species that makes an abundant provision of *honey*.

In the book of Joshua we are informed[2] that God, by means of some animal of this genus, drove out the two kings of the Amorites from before the Children of Israel. In the second volume of Lieut. Holman's Travels—in whom the loss of sight has been compensated by a wonderful acuteness of mental vision—the following anecdote is related illustrative of this fact.[3]

" Eight miles from Grandie ———, the muleteers suddenly called out ' Marambundas, Marambundas !' which indicated the approach of a host of *wasps*. In a moment all the animals, whether loaded or otherwise, laid down on their backs, kicking most violently ; while the blacks, and all persons not already attacked, ran away in different directions, all being careful, by a wide sweep, to avoid the swarms of tormentors that come forward like a cloud. I never witnessed a panic so sudden and complete, and really believe that the bursting of a water-spout

---

[1] See *Introd to Ent*. i. 501.   [2] xxiv. 12.
[3] Quoted in *Lit. Gazette*, Jan. 3, 1835, p. 4.

could hardly have produced more commotion. However it must be confessed that the alarm was not without a good reason, for so severe is the torture inflicted by these pigmy assailants, that the bravest travellers are not ashamed to fly the instant they perceive the terrific host approaching, which is of no uncommon occurrence on the Campos."

I shall now turn to those admirable creatures, which though, as a wise man observes, *they are little among such as fly, their fruit is the chief of sweet things,*[1] those Heaven-instructed mathematicians, who before any geometer could calculate under what form a cell would occupy the least space without diminishing its capacity, and before any chemist existed to discover how wax might be elaborated from vegetable sweets, instructed by the Fountain of Wisdom, had built their hexagonal cells of that pure material, had closed them at the bottom with three rhomboidal pieces, and were enabled, without study, so to construct the opposite story of combs, that each of these rhomboids should form one of those of three opposed cells,[2] thus giving strength to the structure that, in no other place, could have been given to it. Wise in their government, diligent and active in their employments, devoted to their young and to their queen, they read a lec-

---

[1] *Ecclus.* xi. 3.    [2] PLATE XI. FIG. 3.

ture to mankind that exemplifies their Oriental name—*she that speaketh.*[1] Whoever examines their external structure, as has been before observed,[2] will find every part adapted to their various employments.

These valued animals, so worthy of the attention of the sage, as well as the culture of the economist, are almost the only ones of the Order that are guilty of no spoliation, and injure no one: they take what impoverishes none, while it enriches them and us also, by the valuable products which are derived from their skill and labour—true emblems of honest industry.

I shall merely mention the humble-bee,[3] and their subterranean habitations, which are of a much ruder architecture than those of the hive-bee: the cells, however, are made of a coarse kind of wax, but placed very confusedly, nor exhibiting the geometrical precision observable in the latter.[4]

I may here observe that all insects of this Order, in their perfect state, imbibe the nectar from the flowers, but none, the hive and humble bees and one species of wasp excepted, with the view of storing it up for future use.

---

[1] Heb. דבורה

[2] See above, p. 187, and *Introd. to Ent.* i. 481—497, and ii. Lett. xix. xx.

[3] *Ibid,* Lett. xviii.　　　[4] See *Linn. Trans.* vi. t. xxvii.

The last Hymenopterous tribe[1] includes the *ants*, and is almost equally interesting with the preceding one, for the wonderful industry of the animals just mentioned. They are universal collectors; every thing that comes in their way, whether animal or vegetable, living or dead, answers their purpose; and the paths to their nests are always darkened with the busy crowds that are moving to and fro. Their great function seems to be to remove every thing that appears to be out of its place, and cannot go about its own business. I have seen several of them dragging a half-dead snake, about the size of a goose-quill. They do not, however, like the bees, usually store up provisions, but they will imbibe sweet juices from fruits and also from the plant-lice, which may be called their cows.[2] However, almost all their cares and labour are connected with the nurture and sustenance of their young.

I am indebted to the kindness of Lieutenant-Colonel Sykes, of the Bombay army—well known for the zeal and ability with which he investigated the animal productions of the western provinces of India—for some interesting observations upon three species of ants, particularly one, which, from making its nests on the

---

[1] *Heterogyna.* Latr. See *Introd. to Ent.* i. 476—481. ii. Lett. xvii.

[2] *Ibid*, ii. 87—91.

branches of trees, is called the *Tree-ant*, singularly exemplifying the extraordinary instincts of these laborious and provident insects, and which I have his permission to insert in this work.

The *Tree-ant*[1] inhabits the Western Ghauts, in the collectorate of Poona, in the Deccan, at an elevation of from 2,000 to 4,000 feet from the level of the sea. It is of a ferruginous colour, two-tenths of an inch in length ; head of the neuter disproportionably large ;[2] the thorax is armed posteriorly with two sharp spines. When moving the insect turns the abdomen back over the thorax,[3] and the knotty pedicle lies in a groove between the spines. The male is without the spines.[4]

These ants are remarkable for forming their nests,[5] called by the Marattas *moongeeara*, on the boughs of trees of different kinds; and their construction is singular, both for the material and the architecture, and is indicative of admirable foresight and contrivance : in shape they vary from globular to oblong, the longest diameter being about ten inches, and the shortest eight. The nests consist of a multitude of thin leaves of *cow-dung*, imbricated like tiles upon a house, the upper leaf formed of one unbroken sheet, covering the summit like a skull-cap. The

[1] *Myrmica Kirbii.* Sykes.  
[2] PLATE XI. c. FIG. 1, 3.  
[3] PLATE XI. FIG. 3.  
[4] Ibid. Fig. 2.  
[5] Ibid. FIG. 4.

leaves are placed one upon another, in a wavy
or scalloped manner, so that numerous little
arched entrances are left, and yet the interior is
perfectly secured from rain. They are usually
attached near the extremity of a branch, and
some of the twigs pass through the nest. A vertical section presents a number of irregular
cells, formed by the same process as the exterior. Towards the interior the cells are more
capacious than those removed from the centre,
and an occasional dried leaf is taken advantage
of to assist in their formation. The nurseries
for the young broods in different stages of developement are in different parts of the nest. The
cells nearest the centre are filled with very
minute eggs, the youngest members of the community; those more distant, with larger eggs,[1]
mixed with larves; and the most remote, with
pupes near disclosure. In fact, in these last
cells only were found winged insects. The
female is in a large or royal cell, near the centre
of the nest: she is about half an inch long, of
the thickness of a crow-quill, white, and the
abdomen has five or six brown ligatures round it,
like the female of the white ants; the head is
very small, and the legs mere rudiments: she is
kept a close prisoner, and incapable of motion in
her cell, a circumstance in which these appear

---

[1] It should seem from this that the eggs grow.

to approach the white ants, and which indicates that they should form a distinct genus.

There was no store of provisions in the nests; they were indebted therefore for their support to daily labour. We may gain some idea of their perseverance when we consider that the material of which the nest is formed—cow-dung—must have been sought for on the earth, and probably carried from a considerable distance up the trees.

Colonel Sykes related to me another anecdote with regard to an Indian species of ant, which he calls the *large black ant*, instancing, in a wonderful manner, their perseverance in attaining a favourite object, which was witnessed by himself, his lady, and his whole household. When resident at Poona, the dessert, consisting of fruits, cakes, and various preserves, always remained upon a small side-table, in a verandah of the dining-room. To guard against inroads the legs of the table were immersed in four basins filled with water; it was removed an inch from the wall, and, to keep off dust through open windows, was covered with a table-cloth. At first the ants did not attempt to cross the water, but as the strait was very narrow, from an inch to an inch and a half, and the sweets very tempting, they appear at length to have braved all risks, to have committed themselves to the deep, to have scrambled across the channel, and to have reached the object of their desires, for hundreds we found every morning revelling in enjoyment: daily vengeance was executed upon them without lessening their numbers; at last the legs of the table were painted, just above the water, with a circle of turpentine. This at first seemed to prove an effectual barrier, and for some days the sweets were unmolested, after which they were again attacked by these resolute plunderers; but how they got at them seemed totally unaccountable, till Col. Sykes, who often passed the table, was surprised to see an ant drop from the wall, about a foot above the table, upon the cloth that covered it; another and another succeeded. So that though the turpentine and the distance from the wall appeared effectual barriers, still the resources of the animal, when determined to carry its point, were not exhausted, and by ascending the wall to a certain height, with a slight effort against it, in falling it managed to land in safety upon the table. Col. Sykes asks,—is this instinct? I should answer, no: the animal's appetite is greatly excited, its scent probably informs it where it must seek the object of its desire; it first attempts the nearest road; when this is barricaded it naturally ascends the walls near which the table was placed, and so succeeds by casting itself down,—all the while under the guidance of its senses.[1]

It is observed, in the *Introduction to Entomology*, that though ants, " during the cold winters, in this country, remain in a state of torpidity, and have no need of food, yet in warmer regions, during the rainy seasons, when they are probably confined to their nests, a store of provisions may be necessary for them.[2]  Now, though the rainy season, at least in America, as has been stated on a former occasion,[3] is a season in which insects are full of life, yet the observation, that ants may store up provisions in warm countries, is confirmed by an account sent me by Col. Sykes, with respect to another species which appears to belong to the same genus as the celebrated *ant of visitation*,[4] by which the houses of the inhabitants of Surinam were said to be cleared periodically of their cock-roaches, mice, and even rats.[5]  The present species has been named by Mr. Hope, the *provident ant*.[6]  These ants, after long continued rains during the monsoon, were found to bring up and lay on the surface of the earth, on a fine day, its stores of grass seeds, and grains of Guinea corn, for the purpose of drying them. Many scores of these hoards were frequently ob-

---

[1] See above, p. 239, 278, and *Introd. to Ent.* ii. 62.
[2] *Ibid.* 46.    [3] See above, p. 250.    [4] *Atta cephalotes.*
[5] *De Geer.* iii. 607.        [6] *A. providens.*

servable on the extensive Parade at Poona. This account clearly proves that, where the climate and their circumstances require it, these industrious creatures do store up provisions.

From these very interesting communications we may remark how the functions of animals are varied, the same function being often given in charge to tribes perfectly different in different climates. In temperate regions, the principal agents in disinfecting the air by devouring or removing excrement, belong to the Order of *beetles*, but in India, where probably more hands are wanted to effect this purpose of Providence, the *tree-ants* are called in to aid the beetles, by building their nests of this fetid mortar, and thus clear the surface of innumerable nuisances, which probably soon dry and become scentless. In Europe, again, no ants are found to verify Solomon's observation, literally interpreted, but in India we see, and probably it may also be the case in Palestine, provision for the future is not stored up solely by the bees, but the ants, where it is necessary, are gifted with the same admirable instinct.

A circumstance here requires notice, which is almost peculiar to the gregarious Hymenoptera dwelling in a common habitation; in all their communities, besides one or more *prolific* females and males, there is an order of sterile females, which have no connexion with the other sex, and are solely employed in labours and pursuits beneficial to the community at large to which they belong, especially the care and nurture of the young.

The wisdom and beneficial effects of this law, by which the Creator has regulated their communities, and prescribed to all their duties and functions, must be evident to every one. It sets free the majority of the community to give their whole attention to those labours upon which the welfare and existence of their several associations depend. Indeed, if they were all to be prolific, their societies would soon be dissolved, or destroyed by the evils attendant upon an overabundant population; or their increase would be so rapid, that the whole earth would soon be covered by them, to the great annoyance, if not destruction, of the rest of its inhabitants.

Now I am upon this subject, I may add a few remarks upon the kindred societies of *white-ants*, which, though they belong to a different Order, are, in many respects, analogous to those of the true *ants*; and the differences observable between them arise from a marked diversity in the nature of their metamorphosis; namely, that in the last named insects, both larves and pupes are incapable of locomotion, and all the labours of the society, as well as its defence and the care and nurture of the young, are devolved upon a description of its members that are not gifted with the faculty of reproduction : whereas, in the former, the white ants, the larves and pupes, in conformity to the law which, in this respect, regulates the Class to which they belong, are locomotive and more active in those states than in the last or reproductive one, and are therefore fully qualified to act in all the working departments, and to transact the general business of the society ; but as this, in their case, required a conformation of the head and oral organs inconsistent with their use as offensive weapons, another order was necessary to act as sentinels, and to be entrusted with the defence of the nest or termitary, as it is called, and its inhabitants. That such an order exists, we learn from the statements of Smeathman and Latreille, who, both of them, had means of personal investigation, and the latter of whom brought to the investigation the deepest insight into his subject, and the most extensive knowledge of insects and their history possessed by any man in Europe. Upon the accuracy of his statements, therefore, the most entire reliance may be placed. The species[1] he investigated was discovered by himself, in the neighbourhood of Bordeaux, inhabiting the trunks of firs and oaks, immediately under the bark, where, without attacking the bark itself, they formed a great number of holes and irregular galleries. In these societies he

[1] *Termes lucifuga.*

discovered, at all times, *two* kinds of individuals, which were without wings, elongated, soft, of a yellowish white, with their head, trunk, and abdomen distinct; they were active, furnished with six legs, their head large, and the eyes very small, or altogether wanting; but, in one of these kinds of individuals, which compose the bulk of the society, the head is rounded and the mandibles not extended; while in the others, which form not more than one twenty-fifth of the population, the head is much larger, elongated, and cylindrical, and terminated by mandibles that extend from it and cross each other; these Latreille always found stationed at the entrance of the cavities where the others were assembled in greatest numbers : towards the end of the winter and in the spring, he discovered individuals exactly resembling those first mentioned, but having the rudiments of four wings, and in June, the same individuals had acquired four ample wings, had become of a blackish colour, and consisted of males and females; a month later a few only were found in the termitary, which had lost their wings, and eggs now begun to appear laid up in certain labyrinths of the wood.[1]

It is clear from this account that those with a round head and short mandibles are larves,

---

[1] Latreille in *N. D. D'H. N.* xxxiii. 90.

which go through the usual metamorphosis of their tribe, not changing their form, but acquiring wings, first packed up in cases, and afterwards developed. The second description, with the elongated head and crossed mandibles, never acquired wings, and therefore correspond precisely with the neuters amongst ants, only as Providence always economizes means, and wills that nothing be lost or wasted, he has decreed that these locomotive larves and pupes should not live in idleness.

*Order* 7.—We now come to an Order, taking their food by suction, which appear to have been formed to deck our fields and groves with various beauty; but which in their first. state, when they masticate their food, they mar and destroy, often stripping the trees of their leaves, and covering our hedges with their webs full of crawling myriads of devastators. It will be seen that I am speaking of the *Lepidopterous* Order, consisting of three great phalanxes, the *diurnal* fliers, or butterflies, [1] the *crepuscular* fliers, or hawkmoths,[2] the *nocturnal* fliers, or moths,[3] each divided into several genera. Their caterpillars most generally feed upon the foliage of vegetables of every description; but those of some of the lower tribes[4] of moths devour animal substance, such as wool, fur, leather, grease, and the

---

[1] *Papilio.* L.  [2] *Sphinx.* L.  [3] *Phalæna.* L.  [4] *Tineidæ.*

like; some even enter the bee-hive and devour the combs, others the cabinet of the entomologist to prey upon his insects, others even attack the books of the scholar. Their office seems to be to keep in check too luxuriant vegetation, and, in many of the latter instances, the removing of dead animal matter, and every thing putrescent from the surface of the globe.

But this is not the whole, they likewise help to maintain, as has been before observed,[1] half the birds of the air, forming a principal portion of their food; and in some countries, as well as the locusts and white ants,[2] they are eagerly devoured by man himself. There is a certain mountain, in New Holland, as we are informed by Mr. Bennett,[3] called Bugong mountain, from multitudes of small moths, called *Bugong* by the natives, which congregate at certain times, upon masses of granite, on this mountain. The months of November, December, and January, are quite a season of festivity amongst these people, who assemble from every quarter to collect these moths. They are stated also to form the principal summer food of those who inhabit to the south of the snow mountains. To collect these moths, or rather butterflies,[4] the natives make smothered fires under the rocks on which they congregate; and suffocating them with smoke,

---

[1] See above, p. 26.    [2] *Introd. to Ent.* i. 303, 307.
[3] *Wanderings*, &c. i. 265.    [4] *Euplœa hamata.* M'L.

collect them by bushels, and then bake them by placing them on heated ground. Thus they separate from them the down and the wings, they are then pounded and formed into cakes resembling lumps of fat, and often smoked, which preserves them for some time. When accustomed to this diet they thrive and fatten exceedingly upon it.[1] Millions of these animals were observed also, on the coast of New Holland, both by Captains Cook and King.[2] Thus has a kind Providence provided an abundant supply of food for a race that, subsisting solely by hunting or fishing, must often be reduced to great straits.

*Orders* 3 and 11.—The masticating tribe, which present the most striking analogy to the scaly-winged lepidopterous insects, is one of very different habits; mostly bold, rapacious, and sanguinary, they are perpetually chasing other insects, and devouring them, and this they do, not in one, but in all their states. I am speaking here of the Neuropterous Order, especially the dragon flies, those insects of vigorous wing and indomitable force. Every one who compares these with the Heliconian butterflies, the wings of which are sometimes, more or less, denuded of their scales,[3] will perceive that they are

---

[1] Bennett, *ubi supr.* 271.          [2] *Ibid*, 209, note *.
[3] E. G. *Heliconius Quirina, Hippodamia*, &c.

analogues of each other; and one of this Order, the *Ascalaphus*, resembles a butterfly so strikingly, both by its wings and antennæ, that it has been described as one by a very eminent entomologist.[1] The Antlions, and lace-winged flies, in the port of their wings, resemble several moths; and the *Trichoptera*, an osculant Order, but still reckoned amongst the *Neuroptera* by Latreille, in its habit of clothing itself with a case made of various articles, imitate the clothes-moth, and others of that tribe, which invest themselves with cases made of wool, fur, and similar materials.

The dragon-flies in their two first states, by means of their wonderful mask,[2] destroy a vast number of aquatic insects, and in their last an equal number in the air.

The *white ants*,[3] and some kindred insects, like the ants devour every thing but metal, that is exposed to their attacks, particularly timber. A deserted African village is soon removed by them, working under their covered ways; and, in tropical regions, a forest quickly springs up where a busy population ran to and fro a few years before. So that they are amongst the instruments in the hand of Providence, that the places deserted by man shall be restored again to the vegetable and animal races that were in

---

[1] *Scopoli,* see *N. D. D'H. N.* ii. 580.
[2] *Introd. to Ent.* iii. 125.          [3] *Termes.*

possession before he cleared it for his own habitation. The white ants seem to connect this Order with the Hymenoptera by means of the common ants; which, however, as Colonel Sykes informs me, bear the most rooted enmity to them, and destroy them without mercy. In digging up some white ants' nests, in his garden at Poona, he once found *two* queens in one cell, a remarkable anomaly in their history. In the course of the present year I received a letter signed P. T. Baddeley, inclosing a drawing and specimens, of a singular species of white ant, with a head precisely resembling that of an elephant, except that there was no representation of the tusks. The head, which is enormously large compared with the size of the animal, terminates in a long proboscis. Mr. Baddeley found it in great numbers about two years ago, under some teak timber; the only circumstance which he mentions of its habits.

*Orders* 8 and 9.—There are two Orders taking their food by suction, the *Homoptera* and *Hemiptera*, which perhaps should rather be regarded as *Sub-orders*, as Latreille considers them, and which were included by Linné in the same Order with the *Orthoptera* of modern entomologists, to which, in fact, they are contrasted more or less. I shall therefore consider them together.

The *Homoptera* are herbivorous, sucking the

sap of trees and plants,[1] and the principal tribe
of them was celebrated of old, both by Grecian
and Roman bards, under the names of *Tettix*
and *Cicada*, for the far-resounding song of its
males.

This Order contains some of the most singular
monstrosities that the insect world produces;
animals armed with strange appendages and
horns, which in the majority, are processes of
the *trunk*; but, in the *lanthorn-flies*, of the *head*:
the latter have been regarded, as their name
imports, as a kind of lanthorn, given to the
animal to afford it light; but considerable doubt
has been thrown upon the fact. The use of the
arms and processes of the trunk, which are
found chiefly in the male, as well as in many
male Lamellicorn beetles,[2] has not been satisfactorily ascertained; but probably, like the horns
of quadrupeds, and the spurs of male gallinaceous birds, they use them in their mutual
battles.

One of these animals, as producing the *manna*
of the Pharmacopeia, may be regarded as of
some use to mankind. And perhaps, in general,
the tribe, in their perfect state, in which they
imbibe the juice of plants and trees, if not too
numerous, are probably of use to trees that are
over vigorous, and full of sap. In their grub

---

[1] *Phytomyza,* plant-suckers.

[2] *Dynastes, Onthophagus, Copris,* &c.

state, in America, they are very injurious to timber, and fruit trees, into which they introduce their eggs by a remarkable organ or ovipositor.

The proper *Hemiptera*, so called because their wing-covers at the base are of a substance resembling horn or leather, and are membranous at the tip, form the *last* suctorious Order; they are carnivorous, or more properly, *animal*-suckers;[1] for though many of them are found on particular trees and plants, it is not the juices of these that they usually imbibe, but those of the insects that frequent them; there is one, however, too well known in this country, the *bed-bug*,[2] which is more ambitious, extending its attacks, like the flea, to the higher animals, being often found upon pigeons, upon rabbits, and more commonly infesting man himself, during his hours of repose. This Sub-order also presents a great variety of forms, and the bite of some is very venomous.

The *functions* of these are similar to those of other Insects, that derive their nutriment from the higher animals by sucking the blood or juices; but the bugs, being generally *Insect*-suckers, with their juices also suck away their lives, and so are employed to diminish their numbers. The *water-bugs*[3] attack other aquatic

---

[1] *Zoomyza.*          [2] *Cimex lectularius.*

[3] *Hydrometra, Notonecta, Nepa,* &c.

animals as well as Insects, such as fishes, Molluscans, &c.

*Order* 12.—The Orders that are placed as parallels to the *Homoptera* and *Hemiptera*, are the *Orthoptera* and *Coleoptera*. The former includes within its limits Insects of various habits, which may be divided, respect being had to their *food*, into *three* tribes :—those that are *herbivorous*, those that are *carnivorous*, and those that are *omnivorous*.

The *first* of these tribes includes all those Insects known by the common name of *grasshoppers*, and *locusts*;[1] several of those whose wing-covers and wings resemble leaves or flowers;[2] besides other kinds, which I need not mention. The ravages of those first mentioned, especially the locusts, are so well known,[3] that I shall not enlarge upon them.

The *second* tribe consists of what, from the posture they assume, have been called *praying-insects*,[4] some of which also resemble leaves. These are as ferocious and cruel as any of the insect tribes.[5]

The *last* tribe consists principally of the *crickets*[6] and *cock-roaches*,[7] animals that make their ap-

---

[1] *Locusta.*     [2] *Pterophylla.* Stoll. *Saut.* t. i. 3.
[3] See Vol. I. p. 89.     [4] *Mantis. Phyllium.*
[5] *Introd. to Ent.* i. 278.     [6] *Gryllus. Gryllotalpa,* &c.
[7] *Blatta.*

pearance only in the *night*, and feed both on animal and vegetable substances. It has been suggested to me by an eminent and learned Prelate, that the Egyptian plague of *flies*, which is usually supposed to have been either a mixture of different species, or a fly then called the *dog-fly*,[1] but which is not now known, was a *cock-roach*. His Lordship did not assign the reason that led him to adopt this opinion, but the Hebrew name[2] of the animal, which is the same by which the raven also is distinguished, furnishes no slight argument in favour of it. The same word also signifies the *evening*. Now the cock-roach at this time found in Egypt[3] is *black*, with the anterior margin of the thorax white, and they never emerge from their hiding places till the *evening*, both of which circumstances would furnish a reason to the name given it; and it might be called the *evening* Insect, both from its colour and the time of its appearance.

There appears to be a striking analogical resemblance between the bulk of the *Orthoptera* and *Homoptera* to the *Reptiles*, particularly the *Batrachian ;* their leaping and song are the principal points in which they agree, whence the members of the latter Sub-order have usually been called *frog*-hoppers, but in some of the grass-hopper tribe there is also a singular coincidence in their form.[3]

---

[1] Gr. κυνομυια.     [2] רעב     [3] Stoll. *Saut. t.* viii. b. *f.* 29.

*Order* 4.—The *earwigs*[1] form a truly *osculant* Order, between the *Orthoptera* and *Coleoptera*, and partaking of the characters of both, but their habits are so well known that it is not necessary to dwell upon them.

*Order* 13.—Of all the insect Orders which God has created and employed to work his will upon earth, by removing whatever deforms or defiles the face of nature, there is none more remarkable, both for its numbers, the diversities of form and aspect that it exhibits, and of armour both defensive and offensive, and also of its organs of various kinds, and for various uses, than that of which I am now, in the last place, to give some account, the *beetles*, namely, forming the Order *Coleoptera*.

The parallel to this Order amongst the suctorious insects, appears to be the *Hemiptera* Suborder, the wing-covers of some of which,[2] having scarcely any membrane at their extremity, represent the elytra of the Order in question ; indeed the substance of the base of these organs, in the generality, also corresponds with that of the beetles.

Of all the mandibulate Orders there is none that appears to have so universal an action upon every substance, both vegetable and animal,

---

[1] *Forficula.*          [2] *Lygæus apterus, brevipennis, &c.*

both living and dead, as the one before us, but it is difficult to class them according to their food without breaking up natural groups; thus in the great tribe of Lamellicorn beetles, forming Linné's genus *Scarabæus*, we find insects that feed upon a great variety of vegetable food, both liquid and solid; green and putrescent; the feces of animals; and in a few instances, on their flesh.

A very considerable number of this Order are *predaceous* in their habits, and devour without pity, any small animal they can seize and over-power. Of this description is the whole tribe of *ground-beetles*, called by old writers clocks and dors, considered by Linné as forming one genus,[1] but now divided into more than a hundred.

One of the most remarkable of this tribe is the spectre-beetle[2] described by Hägenbach, which is found both in Java and China. In its general aspect, though evidently belonging to the Cara-bidans, it seems to represent the praying-insects, and the spectres;[3] and, from its great flatness, it probably insinuates itself into close places, either for concealment or to lie in wait for its prey.

The splendid tribe of *tiger-beetles*,[4] as they indicate by their fearful jaws, have the same

---

[1] *Carabus.*
[2] *Mormolyce.* PLATE XI. FIG. 1.
[3] *Phasma.*
[4] *Cicindela, Manticora.*

habits, adding a swift flight to the rapid motions on foot which distinguish the other. The grubs of these emulate spiders in some respects, lying in wait for their prey in burrows in which they curiously suspend themselves.[1]  In the waters a considerable tribe of Beetles pursue various aquatic insects, and by means of their oary hind legs swim very swiftly, often suspending themselves at the surface by their anal extremity, near which are two large spiracles for respiration, for they do not respire the water like fishes and the grubs of Dragon-flies. Their larves are armed with tremendous sickle-shaped jaws, through which they pump the juices from fishes as well as insects.

Besides those that are indiscriminate devourers, others confine themselves to particular tribes or species. Thus one of the most splendid of the, so called, *ground-beetles*, named the *sycophant*,[2] ascends the trees and shrubs after the caterpillars which are its destined food, and probably other species of the genus have the same commission. The *rove-beetles*[3] bury themselves in excrement in order to devour the grubs that frequent it. I have before mentioned[4] the wasp-beetle; there are others which, in the same way, attack those of the hive and other bees.[5]

---

[1] *Introd. to Ent.* iii. 152.     [2] *Calosoma Sycophanta.*
[3] *Staphylinus.* L.              [4] See above, p. 334.
[5] *Clerus apiarius,* and *alvearius.*

'Another has a more remarkable instinct, by which it is impelled to seek its nutriment in the slimy snail.[1] There is an insect much resembling a bird-louse that is parasitic on wild bees, which has been thought to be produced from the eggs of the great oil-beetle,[2] but some doubt still hangs on the fact.[3]

Another tribe of beetles have a different commission from their Creator, and instead of *living* ones, feed upon *dead* animals, of every description. To this tribe belong the burying beetles, long celebrated for the manner in which they bury pieces of flesh to which they have committed an egg ;[4] other carrion beetles[5] may be found in considerable numbers of various species and kinds, under every carcass ;[6] even *bones*, after they are denuded of the flesh, are attended by certain insects of this Order, by whose efforts they are completely stripped of every remnant of muscle.[7] Some even find their nutriment in the interior of horns.[8]

Lacordaire observes that the carcasses dry so rapidly in South America, that few necrophagous insects are found there: and that even in the Pampas, and at Buenos Ayres, where animals

---

[1] *Cochleoctonus.*          [2] *Meloe.*
[3] See *Introd. to Ent.* iii. 162. note 6.
[4] *Introd. to Ent.* i. 352.          [5] *Silpha.* L.
[6] *Dermestes. Byrrhus,* &c.
[7] *Nitidula,* &c.          [8] *Trox.*

decompose as in Europe, there are but few of these insects: but their place is supplied by innumerable birds of prey.   As soon as an animal is killed, they fly in crowds from every part of the horizon, though one before was not to be seen.   The most destructive beetles in these countries are those that attack *leather* or skins. Two species of the same genus[1] commit dreadful ravages in the magazines of this article: and in spite of the constant pains that are bestowed to get rid of these insects and their grubs, great losses are suffered.

Another unsightly substance is removed by numberless beetles, whose office is that of scavengers; the celebrated *Scarabæus* of the Egyptians,[2] the symbol, as it is supposed, of the sun, is of this description; the pill-beetle also,[3] equal in fame to the burying one, for trundling its pills, each containing an egg, with the aid of his co-species: many of a smaller type are likewise devoted to the same office.[4]

It is worthy of remark that all these feed only on the excrement of *herbivorous* animals; none having been recorded, I believe, that feed on that of *carnivorous* ones, except a single species[5]

[1] *Dermestes cadaverinus et vulpinus.*
[2] *Scarabæus sacer.*
[3] *Ateuchus pilularius.   Introd. to Ent.* i. 351.
[4] *Sphæridium,* &c.
[5] *Hybosorus geminatus.*

that inhabits *human* excrement solely, but forms no burrow under it.

Others of the order make a transition to the *vegetable* kingdom, by attacking various kinds of fungi, as agarics, Boleti, puff-balls, and the like, which in fact seem to exhibit, in their substance, some analogy to *flesh*. Fabricius has given the name of *Agaric-eater*[1] to a genus that is chiefly found in the *Boletus;* another beetle, however, devours agarics, and is found, I believe, in no other fungus ;[2] and the puff-ball affords a favourite nutriment to others.[3]

Some beetles, or tribes of beetles, are both predaceous, carnivorous, coprophagous, and fungivorous. The Histers will devour carrion, dung, funguses, and putrescent wood : I once found the autumnal dung-beetle[4] in considerable numbers in a dead bird, and Lacordaire mentions others that are carnivorous : he says that the habits of *Trox* approach those of the necrophagous beetles, it being always found under half-dried carcasses, of which they gnaw the tendinous parts. It is found also in the excrements of man and herbivorous animals. *Phanœus Milon* he observed principally under putrescent fishes on the shores of the River Plate.[5]

---

[1] *Mycetophagus. Boletaria.* Marsh.

[2] *Oxyporus maxillosus.*  [3] *Lycoperdina.*

[4] *Geotrupes autumnalis.*

[5] *Ann. des Sc. Nat.* xx. 263. 265.

We have thus had a regular transition, with regard to their food, leading the beetle tribes through the animal to the vegetable world.

Vegetable feeders are innumerable amongst them, the gold,[1] tortoise,[2] and flea beetles[3] all devour plants in both their active states, and some of these are extremely injurious to the farmer[4] and gardener. Many are destructive to seeds, fruits, and roots, numbers of the weevil tribe, and all the Bruchi are of this description.[5]

But of all the beetle tribes the *timber-devourers* are the most numerous ; one of the most splendid and brilliant of the whole Order, the *Bupres-tidans*, belongs to this department, and the still more numerous and more varied *Capricorn* beetles,[6] though less refulgent with metallic splendour, add a vast momentum in the interminable forests of tropical regions, and must be of the greatest use in gradually reducing trees that have been uprooted by tornadoes, or any other cause, to a state of putridity, and finally to dust. Other beetles, of smaller dimensions, and of a cylindrical form, which take their station between the bark and the wood, are instrumental in separating them so as to let in the wet,[7] and

---

[1] *Chrysomela*, &c.    [2] *Cassida*.    [3] *Haltica*.
[4] *Introd. to Ent.* i. 187. 207.    [5] *Ibid*, 172. 176, &c.
[6] *Cerambyx*. L.    [7] *Introd. to Ent.* i. 235. 260.

expose the timber more effectually to the action of the elements.

The great majority, indeed, of this interesting Order derive their nutriment, in their first and last states, from the vegetable kingdom. The Lamellicorns afford a conspicuous instance of this. Even those of them that are coprophagous, feed upon vegetable detritus in some degree animalized; and some are stated to feed indifferently both on excrement and leaves.[1] The giants of the Order, the mighty Dynastidans,[2] appear to feed upon putrescent timber, burrowing in it as well as in the earth. The Melolonthidans, in their first state, devour the *roots* of grass, &c., whence one of the modern genera into which they are divided is named the *root-eater;*[3] in their perfect state, they emerge from their subterranean dwellings, and attack the *leaves* of trees and shrubs, and are sometimes very injurious to them. Again, there are others, which, as it were, disdaining such coarse food, devour the *blossoms* themselves, whence Latreille calls them *Anthobians:* and lastly, the lovely tribe of Cetoniadans, to which the rose-beetle[4] belongs, imbibe the *nectar* of the flowers they frequent.

Many of the weevil tribes are very destructive

---

[1] Lacordaire, *Ann. des Sc. Nat.* xx. 260.

[2] *Dynastes.* M'Leay.　　　　　[3] *Rhizotrogus.*

[4] *Cetonia aurata.*

to stored grain;[1] and others equally so to certain fruits.[2]

Though the *Hymenoptera* and *Neuroptera* Orders are most celebrated for the *associations* which certain tribes instinctively form, this principle does not act in them solely, other Insects have their swarms at certain seasons, as in the case of the New Holland butterflies before noticed; and the beetles afford several instances of it. About the time of the summer solstice, the solstitial beetle[3] may be seen and heard buzzing in vast numbers over the trees and hedges, and a little earlier the cockchafer[4] does the same, and many others of the same family.[5] Lacordaire observed, in Brazil, that two species of *diamond beetles*[6] clustered so on some kinds of Mimosa, that the branches bent under the weight of their glittering burthen.[7]

The same author mentions a curious distinction between the luminosity of the glow-worms and fire-flies in Brazil, which has been confirmed to me by a gentleman sometime resident in that country. In the former, he says, the light perpetually scintillates, but in the latter it is

---

[1] *Calandra.*                    [2] *Cordylia Palmarum.*

[3] *Rhizotrogus solstitialis.*     [4] *Melolontha vulgaris.*

[5] *Hoplia,* &c.            [6] *Entimus imperialis,* and *nobilis.*

[7] *Ann. des Sc. Nat.* xx. 161.

constant;[1] the kind of glow-worm most common in that part of America, belongs to a tribe in which the shield of the thorax does not cover the eyes, and the female is winged as well as the male.[2] Thus in these little illuminators of tropical nights we have a kind of mimic stars and planets, the former of which are so numerous as to fill the air with their scintillations.

The immediate object of this faculty, in these beetles, and in other insects, has not been clearly ascertained; as the females are usually most luminous, it may be to allure the male; or, as most insects fly to the light, it may also bring their prey within their reach; or, again, it may be a defence from their own nocturnal enemies;[3] but whatever be its object with respect to the animals themselves that are gifted with this faculty, they give man an opportunity of glorifying his Creator, not only for the starry heavens, but also for these little flying stars that render night so beautiful and so interesting, where they occur.

In considering the great Class of Insects with reference to their *office*, the first thing that strikes us is their infinite number, not only of

[1] *Ann. des Sc. Nat.* xx. 247.

[2] *In the Introduction to Entomology*, (ii. 407) this genus is named *Pygolampis*, after Aristotle, *Hist. Anim.* l. iv. c. 1.

[3] Vol. I. p. 224.

individuals of the same species, but of different species and even genera, and the vast variety of forms and structures that they necessarily include. When we began the present subject, and, dipping under the waves of ocean, visited the vast world of waters, to survey their various inhabitants; even amongst those that can be seen only by the assisted eye, we saw no traces of such diversity; the number of *individuals*, it is true, were incalculable, but though they have been the objects of research, with so many inquirers, and for so long a period, the number of *species* known fall short of half a thousand, while the number of Insects already in cabinets are stated to be more than two hundred times that number, and even, in our own country, more than *ten thousand* have been enumerated and named.

The momentum of so vast a body of animals, everywhere dispersed, and daily and hourly at work in their several departments, must be incalculable; and this momentum must be doubled by the circumstance that so singularly distinguishes a large proportion of them; I mean that the different periods of their existence are passed under different forms, during which they have quite different functions assigned them, and are fitted with different organs, being, when they are first disclosed from the egg, masticators of solid and grosser food, and in their last state imbibing nectareous fluids. The connection of the first is with the *leaves* of the plant, to them they are committed by the mother as soon as they are extruded from her matrix, and they supply them with their earliest and latest food; but when she is disclosed in all her beauty, dressed as it were in her bridal robes, the connection is between her and the *flower*, her lovely analogue, from them she imbibes the sweet fluid which their nectaries furnish, and now, instead of a devourer, she abstracts merely what is redundant, which, while it contributes to her own enjoyment and support, in the case of the bee, enriches man himself.

We behold, then, this immense army of devourers, varying so infinitely in their instincts, as well as their forms, supplying many animals with the whole of their subsistence, and forming a considerable portion of that of others, and feel convinced that Providence has not placed them in their position, and given them such a variety of organs, except with the view to some great general benefit to those animals amongst whom he has placed them; and this benefit is not so much perhaps the reducing the numbers of their own class within due limits, though that is a most important object, as removing nuisances, which would deform, or in any way infect the earth and its inhabitants. For this the Insect world is principally distinguished as to its functions. It consists of the scavengers of the earth, and the pruners of its too luxuriant productions.

With respect to ornament and pleasurable sensations, which were certainly the object of our beneficent Creator, as well as our profit and utility—next to the birds, nothing adds more to the life of the scene before us, during the diurnal hours, and even sometimes the nocturnal, than the vast variety of insects that are flying, running, and jumping about in all directions, all engaged in their several pursuits,—the bees humming over the flowers ; the butterflies opening and shutting their painted wings to the sun ; the gnats, and gnat-like flies, rising and falling alternately in the sunbeams; the beetle wheeling his droning flight; others coursing over the ground ; the grasshopper chirping in every bank,—all adding to the general harmony, and combining to make the general picture one of life and Love ; and speaking, each in different sort and manner, the praises of its Creator, and calling upon man to join in the general hymn.

## CHAPTER XXI.

### *Functions and Instincts. Fishes.*

THE animals we have hitherto considered have been destitute of an internal jointed vertebral column and its bony appendages; and though some, as the Cephalopods and some slugs,[1] have a kind of internal bone, and in one Order of Polypes[2] the axis is sometimes articulated, yet these, especially in the latter instance, merely indicate an analogical relation, but no affinity. In none of these instances is this internal bone perforated for the passage of a spinal marrow, as in a real vertebrated column; we now, however, enter that superior section of the animal kingdom, the individuals belonging to which, with scarcely any exception, are built upon the column in question, incasing a spinal marrow, and terminated at its upper extremity by a bony casket, calculated to contain and protect the most precious and wonderful of all material substances, the cerebral pulp, by which the organs of sense perceive; the will moves the members; the mind governs the outward frame;

---

[1] VOL. i. p. 305.　　　　[2] Ibid. p. 177.

and, in the king of animals, an immortal spirit, is enabled to seek and secure a higher destiny.

This change in the *structure* of animals was rendered necessary by an increase in their *bulk*, for though there are some of the invertebrated Sub-kingdom, as the fixed Polypes and several of the Cephalopods, that are of as large dimensions, and a few of the vertebrated, as the humming birds,[1] and the harvest mouse,[2] that are not so large as some insects; yet the generality of those distinguished by a vertebral column form a striking contrast, as to magnitude, with those that are not. Besides this, as these animals, by the will of their Creator, were to be endowed as they ascended in the scale, with gradually increasing intellectual faculties, it was necessary that the principal seat of those faculties should be differently organized. A different organ of respiration also, as well as of circulation, in the great body of vertebrates, required an internal cavity defended from the effects of pressure.

Having premised these general observations, we are next to consider what animals form the basis of the vertebrated Sub-kingdom. Most modern zoologists appear to be of opinion that the *Fishes* occupy this position, and, taking all circumstances into consideration, this seems the

---

[1] *Trochilus.*        [2] *Mus messorius.*

station assigned to them by their Creator; still there are characters in some of the *Reptiles* that seem to connect them more immediately with the *Insects*. The metamorphoses, particularly of the Batrachian Order, are of this description; as is likewise the carapace, or shell of the Chelonians, of which the vertebral column and ribs form the basis. Those extraordinary animals, the hag[1] and the lamprey,[2] half worms and half fish, by means of the leech, evidently connect the Fishes with the Annelidans.[3] Perhaps those butterflies of the ocean, the flying fishes,[4] with their painted wing-fins with branching rays, may look towards the *Lepidoptera* amongst Insects, but there is no direct connection at present discovered between the two Classes.

The characters of the Class of Fishes are—*Body* with a vertebral column, covered with *scales*, and moved by *fins*. *Respiration* by permanent *gills*. *Heart* with only one *auricle* and one *ventricle*; *blood* red, cold.

Fishes are distinguished from the other vertebrated animals, especially birds and beasts, by their mode of *respiration*; the *latter* breathing the atmospheric air, are furnished with *lungs*,

---

[1] *Gastrobranchus.* (*Myxine.* L.)   [2] *Pteromyzon.*
[3] Sir E. Home, *Philos. Trans.* 1815, 265.
[4] *Exocœtus volitans*, &c.

which receive that element, oxygenate the blood, and again expel it in a different state; while the *former*, which must decompose the water for respiration, breathe by means of *gills*, found also in many invertebrates; these are usually long, pointed plates, disposed like the plumules of a feather, or teeth of a comb, in fishes attached to bony or cartilaginous bows; each of them, according to Cuvier, covered by a tissue of innumerable blood-vessels; but, according to Dr. Virey,[1] having a minute vein and artery. In the gill of a cod-fish, which I have just examined under a microscope, a vein and artery traverse each plate longitudinally at the margin, which appear to be pectinated, at right angles on each side, with innumerable minute branches, and resemble, in this respect, the gills of Crustaceans.[2] Thus the blood is oxygenated by the air mixed with the water, and carried to the heart, whence it is distributed to the whole body. So that the aërated water produces the same effect upon the blood in the branchial vessels, as the air does upon that in our lungs.

We know, by experience, how soon an animal that breathes by lungs, if it remains only a few minutes under water, and is cut off from all communication with the atmosphere, is suffocated and dies; and that all aquatic animals

---

[1] *N. D. D'H. N.* iv. 330.  [2] Latr. *Cours. D'Ent. t.* 2. *f.* 2.

that have not gills, or something analogous, as all the water-beetles, the larves of gnats, &c. are obliged, at certain intervals, to seek the surface for respiration. Whence we may learn what an admirable contrivance of Divine Wisdom is here presented to us, to enable the infinite host of fishes to breathe as easily in the water as we do in the air.

When we sum up all the diagnostics of the Class we are considering, we can trace, at every step, so that, almost, *he that runs may read*, Infinite Power in the construction, Infinite Wisdom in the contrivance and adaptations, and Infinite Goodness in the end and object of all the various physical laws, and in all the structures and organizations by which they are severally executed, which strike the reflecting mind in this globe of ours. What else could have peopled the waters, and the air, with a set of beings so perfectly and beautifully in contrast with each other, as the fishes and the birds. Sprung originally from the same element, they each move, as it were, in an ocean of their own, and by the aid of similar, though not the same, means. The grosser element they inhabit required a different set of organs to defend, to propel and guide, and to sink and elevate the *fish*, from what were requisite to effect the same purposes for the *bird*, which moves in a rarer and purer medium; yet as both were *fluid* mediums, consisting of the same elements, though differently combined ; analogous organs, though differing in substance, structure, and number, were required.  For what difference is there between swimming and flying, except the element in which these motions take place?  The fish may be said to *fly* in the *water*, and the bird to *swim* in the *air;* but perhaps the movements of the aquatic animal, from its greater flexibility and the number of its motive organs, is more graceful and elegant than those of the aërial.   The *feathers* of the one are analogous to the *scales* of the other ; the *wings* to the pectoral *fins;* and the *tail* of both acts the part of a *rudder*, by which each steers itself through the waves of its own element.

One distinctive character of fishes is taken from the *scales* that cover and protect their soft and flexile forms from injury.   Scales, however, are not peculiar to *fishes*, since many *reptiles*, as the Saurians, and some quadrupeds, as the Pangolin,[1] are armed by them.   Scarcely any species of fish is really without them.   In some, upon which when living they are not discoverable under a microscope, when they are dead, and the skin is dry, scales are readily detected and detached.   These organs vary greatly in form : sometimes they resemble spines, at others they are tuberculated ; but most commonly they

---

[1] *Manis.*

are plates, often carinated, and varying in shape,
some being round, others oval, others again
angular; sometimes also they are finely denticulated. In some fish they are separated, in
others they touch, often so as to form together
the resemblance of a beautiful piece of mosaic,
and in many they are imbricated.[1] In those
that rarely approach the shore, and are exposed
only to slight friction, they are fastened by a
smaller portion of their circumference; but in
in-shore fishes they are more firmly fixed, and
covered partly by the epidermis, which, in
those that live and burrow in the mud, almost
entirely envelopes them. Some fishes set up
their spines like a hedgehog; and most, when
alarmed, seem to have the power of erecting
them more or less. Had we the means of ascertaining the situation and circumstances of every
individual, we should find that, in every case,
the figure and connexion, and substance of the
scales, was ruled by them. A proof of this may
be seen in those fishes whose integument consists of hard scales, united together so as to form
a tesselated coat of mail. I allude to the
*Ostracions*, whose organs of locomotion seem not
calculated to effect their escape when pursued;
the want of speed, however, is compensated by
a covering that the teeth of few of their enemies

[1] Roget, *B. T.* i. 116.

can penetrate : the same remark applies to those fishes that can inflate themselves into a globe,[1] in some of which the fins are so minute, as to be scarcely discoverable. In these the scaly spines, when erected, assist in preventing the attack of enemies.

I have given a detailed account of the *fins* of fishes on a former occasion.[2] I shall therefore here only consider the motions of which they are the organs, and their theatre.

Though the *birds*—if we consider the whole atmosphere of the globe, whether expanded over earth or sea, as their domain—may perhaps have a wider range than the *fishes*, yet when we further consider that, besides the whole extent of the ocean, and the seas in connection with it, with all its unfathomable depths and abysses, and all the rivers that flow into it—all the innumerable lakes also, and other stagnant waters, on mountains, and at every other elevation, that the earth's surface contains, belong to the fishes, and compare at the same time the greatest depth to which they descend with the greatest height to which birds ascend, we may conclude that, with regard to its *extent*, their habitable world may be nearly commensurate with that of their rivals or analogues.

As to their *motions*, in their element, birds of

---

[1] Roget, *B. T.* i. 433.      [2] See above, p. 135.

the most rapid and unwearied wing must yield the palm to them; the eagle to the shark, and the swallow to the herring and salmon. The form of fishes, generally speaking, is particularly calculated for swift and easy motion; and the resistance of the fluid in which they move seems never to impede their progress. While birds that undertake long flights are often obliged to alight upon vessels for some rest and renovation of strength, fishes never seem exhausted by fatigue, and to require no respite or repose. Sharks have been known to keep pace with ships during long voyages; and, like dogs, they will sport round vessels going at several knots an hour, as if they had plenty of spare force.[1] The thunny darts with the rapidity of an arrow, and the herring goes at the rate of sixteen miles per hour. But though many fishes thus pursue an unwearied course without any intervals of repose, yet there are some that often appear to sleep. Inflating its natatory vesicle, our fresh-water shark, the pike, in the heat of the day, rises nearly to the surface, and there remains perfectly motionless and apparently asleep: at this time he is easily snared, by passing a running noose of wire over his tail, and by a sudden jerk bringing him on shore.

The *eye* of fishes is like that of the higher

---

[1] *N. D. D'Hist. Nat.* xxvii. 247.

animals, but of a substance that makes the access of the water to it no more troublesome than that of the air to terrestrial animals. Generally speaking, it is protected by no eyelid or nictitant membrane. One genus, however, removed from the *gobies*,[1] has the *former;* and a species of *bodian*,[2] from the equatorial seas, has a moveable membranous valve above each eye, with which, at will, it can cover it, that seems analogous to the *latter*. The eye of the eel, and other serpentiform fishes, which are usually buried and move about in the mud, is covered, through the provident care of their Creator, by an immoveable membrane; and in several species the organ can be withdrawn to the bottom of the socket, and even concealed, in part, under its margin. But the most singular kind of eye in the Class, and that in which the forethought of the Deity is most conspicuous, is that of the *Anableps,* a viviparous fish, inhabiting the rivers of Surinam, and called by the natives the *four-eyed* fish. If the cornea of this eye be examined attentively it will be found that it is divided into two equal portions, each forming part of an individual sphere, placed one above and the other below, and united by a little narrow membranous, but not diaphanous, band, which is nearly horizontal when the fish is in its natural position; if

[1] *Periophthalmus.*        [2] *B. palpebratus.*

the lower portion be examined, a rather large
iris and pupil will be seen, with a crystalline
humour under it, and a similar one with a still
larger pupil in the upper portion. The object of
Divine Wisdom in this unparallelled structure, if
we may conjecture from the circumstances of the
animal, is to enable it to see near and distant
objects at the same time—the little worms below
it that form its food, with one pupil and iris, and
the great fishes above it or at a distance, which
it may find it expedient to guard against, with
the other.

The senses of smell and hearing have no external avenue in fishes. The former is the most
acute of all their senses. Lacepede says it may
be called their real eye, since by it they can
discover their prey or their enemies at an immense distance; they are directed by it in the
thickest darkness, and the most agitated waves.
The organs of this sense are between the eyes.
The extent of the membranes on which the ol-
factory nerves expand, in a shark twenty-five
feet long, is calculated to be twelve or thirteen
square feet.

The *teeth* of fishes may be divided into the
same kinds as those of quadrupeds; they have
their laniary, incisive, and molary teeth; they
are differently distributed, according to the species and mode of life; some are almost immove-
ably fixed in bony sockets, others in membra-

nous capsules, by which means they can be elevated or depressed at the will of the animal. They not only have often many rows of teeth in their mouth, but even their palate, their throat, and their tongue are sometimes thus armed.[1] And this accumulation of teeth is not confined to the fiercest monsters of the deep, but even some herbivorous fishes have several rows of molary teeth.    An instance of this is afforded by a jaw of some unknown fish, perhaps a Siluridan, in my possession, in which there are six rows of such teeth, the anterior ones being somewhat conical.    This specimen was found on the shore of one of the lakes in Canada, and belonged to a fish, which the friend who gave it to me stated was much relished by the Indians.

Many of the organs of the members of this Class are more independent of each other than those of warm-blooded animals; they seem less connected with common centres, in this respect resembling vegetables, for they may be more materially altered, more desperately wounded, and more completely destroyed, without any mortal effect.    Many of their parts, as the fins, if mutilated, can be reproduced.    Indeed a fish, as well as a reptile, can be cut, torn, or dismembered without appearing to suffer materially. The shark, from which a harpoon has taken a

---

[1] Plate XIII. Fig. 3.

portion of its flesh, pursues his prey with the usual avidity, if his blood has not been too much exhausted. We see in this a merciful provision, that animals so much exposed to injury should suffer less from it than those which are better protected, either by their situation or structure.

Fishes are amongst the most long-lived animals. A pike was taken, in 1754, at Kaiserslautern, which had a ring fastened to the gillcovers, from which it appeared to have been put into the pond of that castle by the order of Frederick the Second, in 1487, a period of two hundred and sixty-seven years. It is described as being nineteen feet long, and weighing three hundred and fifty pounds! !

Though the animals of the Class under consideration are not generally remarkable for their sagacity, yet they are capable of instruction. Lacepede relates that some, which for more than a century had been kept in the basin of the Tuilleries, would come when they were called by their names; and that in many parts of Germany trout, carp, and tench are summoned to their food by the sound of a bell.[1]

At the first blush it seems as if fishes took little care or thought for their offspring; but when we inquire into the subject, we find them assiduous to deposit their eggs in such situations

[1] *Hist. des Poiss. Introd.* cxxx.

as are best calculated to ensure their hatching, and to supply the wants of their young when hatched; but sometimes they go further, and prepare regular *nests* for their young. Two species, called by the Indians, though of different genera,[1] by the name of the *flat-head* and *round-head hassar*, have this instinct, and construct a nest, the former of leaves and the latter of grass, in which they deposit their eggs, and then cover them very carefully; and both sexes, for they are monogamous, watch and defend them till the young come forth. General Hardwicke mentions a parallel instance in the *go-ramy*,[2] of the Isle of France, a fish of the size of the turbot, and superior to it in flavour, cultivated in the ponds of that island.

It has been observed that some fishes, when dead, emit a phosphoric light, I have particularly noticed this in the mackarel, but others do this when living. The *sun-fish*[3] which sometimes has been found of an enormous bulk,[4] when swimming yields a light, which looks like the reflection of the *moon* in the water, whence it has also been called the moon-fish—and the spectator in vain searches for that planet in the heavens. Sometimes many individuals swim together, and

---

[1] *Doras* and *Callicthys.*

[2] *Osphromenus olfax.*          [3] *Mola.*

[4] One is said to have been caught in the Irish sea twenty-five feet long! !—Lacep. *Hist.* 511.

by their multiplied luminous disks, generally at some distance, compose a singular and startling spectacle ; and if we take into consideration the magnitude of these animals,[1] we may conceive the wonder and amazement that would agitate the mind of any one when he first beheld such an army of great lights moving through the waters. For what purpose Providence has gifted the sun-fish with this property, and how it is produced, has not been ascertained. It may either be for defence or illumination.

Few animals, with regard to magnitude, present to the eye such enormous masses as some fishes; leaving the *whales* out of the question, which though aquatic, belong to another Class, what quadruped can compete with the *shark*, which is also a phosphoric fish. That tribe called by the French *Requins*,[2] which is thought to be synonymous with the *Carcharias* of the Greeks, and one of which was probably the sea-monster, mistranslated the *whale*, which swallowed the disobedient prophet—are stated to exceed thirty feet in length; another[3] of a different tribe, is still larger, sometimes extending to the enormous length of more than *forty* feet!![4] Next to the sharks, the *rays*, nearly akin to them, exceed in their magnitude; they are some

---

[1] *Hist. of Waterford*, 271. Borlase, *Cornw.* 267.
[2] *Carcharias*. Cuv.　　　　[3] *Squalus maximus*.
[4] N. D. D'H. N. xxix. 192. xxxii. 74.

times called sea-eagles, because in their rage and
fury they occasionally elevate themselves from
the water, and fall again with such force as to
make the sea foam and thunder.  An individual
of a species[1] of this tribe, called by the sailors the
*sea-devil*, taken at Barbadoes, was so large, as to
require *seven pairs* of oxen to draw it on shore!![2]

If we consider the vast tendency to increase of
the oceanic tribes, that where a terrestrial animal
gives birth to a single individual, a marine one
perhaps produces a *million*, we may conceive
that if no check was provided to keep their
numbers within due limits, they would so fill the
waters as to interfere with each other's and the
general welfare.  The Cod-fish alone, which,
according to Leeuwenhoek and Lacepede,[3] produces more than nine millions of eggs in one
year, if neither man, nor shark nor other predaceous fish, made it their food, would so fill the
ocean in congenial climates, in the course of no
long period of time, that there would scarcely be
space for the motions or life of any other marine animal : the same may be said of almost all
the migratory fishes.  In these circumstances
we see the reason why such enormous monsters
were created that could swallow them by hundreds, why their yawning mouth and throat were

---

[1] *Raia Banksiana.*
[2] Lacep. *Hist. des Poiss.* ii. 116.
[3] Leeuwenh. *Epist.* iii. 188. Lacepe. *Hist.* Ibid. 393.

planted with teeth and fangs of different descriptions, fixed and moveable, arranged in many a fearful row of bristling points, and why this tremendous array has been mustered in the mouth of animals of such never-sated voracity, and of such unmitigated cruelty and ferocity.

Still though the scene is one of blood and slaughter, yet He whose tender mercies are over all his works, has fitted the creatures exposed to it for their lot. Cold-blooded animals, as I lately observed, do not suffer from the various dismemberments to which their situation exposes them, like those of a higher and warmer temperature, whence we may conclude, that great pain and anguish are not felt by them.

Another function of these tremendous animals is to devour all *carcasses*, which, from whatever cause, are floating in the water, thus they act the same part in disinfecting and purifying the ocean, that the hyænas and vultures, their terrestrial analogues, and other animals, do upon earth.

Another lesson may be learned from the existence of these terrible monsters ; for if God fitted them to devour, he fitted them also to instruct. The existence of creatures so evil, and such relentless destroyers of his works in the material world, teach us that there are probably analogous beings in the spiritual world ; and what occasion we have for watchfulness, to escape their destructive fury.

There is nothing more remarkable in the Class

we are considering than the infinite variety and singularity of the figures and shapes of fishes. It has been thought that the ocean contains representatives of every terrestrial and aërial form. However this be, it may be asserted that the forms of fishes are more singular and extraordinary, more grotesque, and monstrous, than those of any other department of the animal kingdom ; but on this subject I need not enlarge.

Having made these general remarks upon fishes, I shall next say something on their *Classification*. Of all the Classes of animals, that of Fishes, as Baron Cuvier observes, is the most difficult to divide into Orders. Linné considered what have been usually denominated *Cartilaginous Fishes*, as forming a section of his *Amphibians :*[1] but the former illustrious naturalist has very judiciously arranged them with the fishes. Ichthyologists in general agree with Cuvier in dividing this Class into two Sub-classes—viz. *Osseans*, in which the skeleton is *bony* and formed of bony *fibres ;* and *Cartilagineans*, in which it is *cartilaginous* and formed of calcareous *grains*. Lacepede, the most eminent of modern Ichthyologists, has observed that there is a striking resemblance or analogy between certain points of these two Sub-classes, of which he has given a table drawn up in a double series, which I shall here subjoin.

[1] *Nantes.*

| CARTILAGINEANS. | OSSEANS. |
|---|---|
| Petromyzon. Gastrobranchus. | Cæcilia. Muræna. Ophis. |
| Raia . . . . . . . . | Pleuronectes. |
| Squalus . . . . . . . | Esox. |
| Accipenser . . . . . . | Loricaria. |
| Syngnathus . . . . . . | Fistularia. |
| Pegasus . . . . . . . | Trigla. |
| Torpedo. Tetrodon . . . . . . | Gymnotus. Silurus. |

Cuvier also remarks, with respect to the animals of the present Class, that they form two distinct *series*,[1] which in another place he says, cannot be considered as either superior or inferior to each other.

Many genera of the Cartilagineans, he thinks, approach the Reptiles by some parts of their organization, whilst it is almost doubtful whether others do not belong to the Invertebrates.[2]  He has made no remark with respect to the connection of the *Osseans* with the above Class : though his thirteenth Family consists of fishes that have always gone by the name of *fishing-frogs*,[3] from the resemblance which they exhibit to that animal, and from their pectoral fins assuming the appearance of legs.[4]  The species of one genus[5] resemble a fish with a lizard on its back, the head being overshadowed by a conical horizontal horn, in the sides of which the eyes are fixed, so that the lower lobe simulates

[1] *Règne Anim.* ii. 128.     [2] *Ibid.* 376.
[3] *Lophius.* L.     [4] PLATE XIII. FIG. 1.
[5] *Malthus.*

the head of a fish, and the upper one that of a lizard.[1] This family of fishes, as well as the *lump-fish*,[2] in his *Lectures on Comparative Anatomy*, Cuvier classed with the Cartilagineans.

It is not to be expected that I should be able to thread my way through a labyrinth, in which this great man confesses himself to be at a loss; and therefore I shall not attempt any alteration of his system, though confessedly the reverse of *natural* with respect to the *Orders* into which he divides it, but leave the subject to an abler hand, M. Agassiz, who is reported to have undertaken it, and in the mean time, give a popular summary of Baron Cuvier's Orders, as I find them in the last edition of the *Règne Animal*.

*Sub-class* 1.—The *Cartilagineans*, which, as allied to the Annelidans, I shall place first, are divided by Cuvier into *three* Orders,[3] viz. the *Cyclostomes*, or suckers; the *Selacians*; and the *Sturionians*.

*Order* 1.—The *Cyclostomes*, or suckers, with regard to their skeletons, are the most imperfect of all the Vertebrates, They have neither *pectoral* nor *ventral* fins. Their *body*, apparently headless and eyeless, terminates anteriorly in a

[1] PLATE XIII. FIG. 2.    [2] *Cyclopterus.*
[3] *Ubi supr.* 128. where Cuvier arranges them in the Order here adopted, but when he gives the details of the Sub-class, he reverses it. Ibid. 378.

circular or semicircular fleshy *lip*, supported by a cartilaginous ring. Their *gills* consist of pouches instead of pectinated organs. By means of their mouth, which, as well as the tongue, is armed with teeth, they fix themselves to fishes, and derive their nutriment from them. The *lamprey*,[1] *lamperne*,[2] and *hag*, &c. belong to this Order.

*Order 2.*—The *Selacians* have gills, fixed by their outer margin, and not disengaged as in the Osseans, and they expel the water by lateral openings. To this Order the *sharks* and the *rays* belong.

*Order 3.*—The *Sturionians* agree with the Ossean Fishes in their gills, but their skeleton is cartilaginous. They have only a single orifice, covered with an operculum. The sole genera included in this Order are the *Sturgeon*[3] and the *Sea-ape*.[4]

*Sub-class 2.*—The *Osseans* Cuvier divides into *four* Orders, viz. *Acanthopterygians, Malacopterygians, Lophobranchians*, and *Plectognathians*. These Orders, for reasons before assigned,[5] I shall reverse.

*Order 1. (Cuv. 6.)*—*Plectognathian* Fishes. *Gill-covers* concealed under a thick skin. *Ribs* rudimental. *Ventral fins* wanting. To this Order

[1] *Petromyzon fluvialis*, &c.    [2] *P. branchialis?*
[3] *Accipenser*.    [4] *Chimæra monstrosa.*
[5] Vol. I. p. 145, and above, p. 320.

belong the *Coat of Mail-fish*,[1] the *Sun-fish*,[2] and the *Bladder-fish*.[3]

*Order* 2. *(Cuv. 5.)*—*Lophobranchian* Fishes. So called because their gills are not pectinated, but disposed in tufts, as is the case likewise with some Annelidans ;[4] body ridged longitudinally, covered with hard scales, united to each other; *mouth* elongated. To this Order belong those singular animals—the *dragonet*,[5] the *horse-head*,[6] or *sea-horse*, and the *sea-needle*.[7]

*Order* 3. *(Cuv. 2.)*—*Malacopterygian*, or soft-rayed Fishes. *Rays* not spiny, except sometimes the first of the dorsal or pectoral fins. This great Order Cuvier divides into three *Orders*, or rather *Sub-orders*, which I shall give inversely.

*Sub-order* 1. *(Cuv. 4.)*—*Apode Malacopterygians*. *Body* serpentiform, elongated; *skin* thick, soft, and slimy. To this Sub-order belong the *common-eel*,[8] the *conger-eel*,[9] and the *electric-eel*,[10] which have many points in common with the cylostomous fishes of the preceding Sub-class, and with respect to their form seem to look both towards the *Annelidans*, and more especially to the *Ophidian* Reptiles.

---

[1] *Ostracion.*     [2] *Mola.*
[3] *Diodon.*     [4] See above, p. 127.
[5] *Pegasus.*     [6] *Hippocampus.*
[7] *Syngnathus.*     [8] *Muræna Anguilla.*
[9] *M. Conger.*     [10] *Gymnotus.*

*Sub-order* 2. *(Cuv.* 3.*)*—The *Sub-brachian Malacopterygians*. *Ventral* fins attached under the *pectoral*. In this Order we find the *sucking-fish*,[1] the *lump-fish*,[2] the *flat-fishes*, and the *cod-fish*,[3] which seems an heterogenous mixture; the flat-fishes seem clearly entitled to rank as an Order.

*Sub-order* 3. *(Cuv.* 2.*)*—*Abdominal Malacop-terygians*. *Ventral* fins attached under the abdomen and behind the *pectoral*. Here, as we ascend, we meet with the *sprat*,[4] the *herring*,[5] the *hassar*,[6] the *salmon*,[7] the *anableps*, the *roach*,[8] *tench*,[9] and *carp*.[10]

*Order* 4. *(Cuv.* 1.*)*—The *Acanthopterygians*, or spiny-rayed Fishes. First rays of the *dorsal* fin, or of the *first* dorsal fin, spiny, or dorsal *spines* in the place of dorsal *fins*. Under this vast Order are arranged an infinity of families and genera, which Cuvier seems to lament that he was obliged to leave together.[11] The *tobacco-pipe-fish*,[12] the *rasor-fish*,[13] the *fishing-frogs*,[14] the *lyre-fish*,[15] the *John Dory*,[16] the *sword-fish*,[17] the

---

[1] *Echeneis.*  [2] *Cyclopterus.*  [3] *Gadus.*
[4] *Clupea Sprattus.*  [5] *C. Harengus.*
[6] *Doras. Callicthys.*  [7] *Salmo.*
[8] *Cyprinus rutilus.*  [9] *C. Tinca.*
[10] *C. Carpio.*  [11] *Règne Anim.* ii. 131.
[12] *Fistularia.*  [13] *Coryphæna.*
[14] *Lophius. Malthus, Batrachus.*
[15] *Callionymus Lyra.*  [16] *Zeus Faber.*
[17] *Xiphias.*

*mackarel*,[1] the *gurnard*,[2] the *mullets*,[3] and the *perch*,[4] are amongst those that belong to this Order.

It is impossible to consider the Orders of Fishes as we have done those of Insects, and give any satisfactory account of the functions and instincts of the several families and tribes that compose them. We cannot dip beneath the waves, to visit the depths of the ocean, that we may investigate their manners and history, but, doubtless, we may conclude, that the same Wisdom, Power, and Goodness, which we find so visibly manifested in the structure and operations of all the animals that are under our eyes and inspection, have equal place and are equally conspicuous, when brought into view, in the marine and other aquatic animals. We know by experience that a large portion of them are of the greatest benefit to mankind, and the rest, from the gigantic shark to the pigmy minnow, each in their place, and engaged in the fulfilment of their several functions, are, we may conclude, equally beneficial, though in a way that we cannot fully appreciate.

I have had more than one occasion to enlarge upon some of those parts of the history of fishes

---

[1] *Scomber Scombrus.*    [2] *Trigla Gurnardus.*
[3] *Mullus.*    [4] *Perca.*

with which we are acquainted,[1] I shall therefore only add here some particulars with respect
to the habits of a few individuals which may
throw some light upon their history.

Amongst the Cyclostomous Cartilagineans the
*hag* is distinguished by a singular means of
escape from its enemies. This animal adheres
to fishes by creating a vacuum by means of
its lips; this effected, it lacerates them with
its teeth, without their being able to shake it off,
and then, like the leech, it sucks their blood and
juices; but since, when thus fixed and employed, it might easily become the prey of
other fishes, Providence has enabled it to conceal itself from them, by means of the excrement which, when in danger, it emits, and
which remains for a time near it, detained by
the slime which exudes from its pores. This
is so abundant that Kalm, having put one in
a large tub of sea water, it became like a clear
transparent glue, from which he could draw
threads, even moving the animal with them.
A second water, upon its being again immersed,
in a quarter of an hour, became the same. Sir
E. Home was of opinion that these animals are
hermaphrodites.

Amongst all the diversified faculties, powers,

[1] VOL. I. 106—124.

and organs, with which Supreme Wisdom has gifted the members of the animal kingdom to defend themselves from their enemies, or to secure for themselves a due supply of food, none are more remarkable than those by which they can give them an electric shock, and arrest them in their course, whether they are assailants or fugitives. That God should arm certain *fishes*, in some sense, with the lightning of the clouds, and enable them thus to employ an element so potent and irresistible, as we do gunpowder, to astound, and smite, and stupify, and kill the inhabitants of the waters, is one of those wonders of an Almighty arm which no terrestrial animal is gifted to exhibit. For though some quadrupeds, as the cat, are known, at certain times, to accumulate the electric fluid in their fur, so as to give a slight shock to the hand that strokes them, it has never been clearly ascertained that they can employ it to arrest or bewilder their prey, so as to prevent their escape. Even man himself, though he can charge his batteries with this element, and again discharge them, has not yet so subjected it to his dominion, as to use it independently of other substances, offensively and defensively, as the electric fishes do.

The fishes hitherto ascertained to possess this power belong to the genera *Tetrodon, Trichi-*

*urus, Malapterurus, Gymnotus,*[1] and *Raia.*[2] The most remarkable are the three last.

The faculty of the *Torpedo* to benumb its prey was known to Aristotle,[3] and Pliny further states,[4] that conscious of its power, it hides itself in the mud, and benumbs the unsuspecting fishes that swim over it. The Arabians, when they cultivated the sciences so successfully, had observed this faculty both in the Torpedo and the Malapterurus, and perceiving an affinity between the electric fluid of the heavens and that of these fishes, called them Raash,[5] a name signifying *thunder.*

The electric organ in the *Malapterurus*[6] extends all round the animal, immediately under the skin, and is formed of a mass of cellular tissue, so condensed and thick as, at first, to look like bacon; closely examined, it is found to consist of tendinous fibres, which are interlaced together, so as to form a net work, the cells of which are filled with a gelatino-albuminous substance, the whole accompanied by a nervous system, differing from that of the *Torpedo* and *Electric-eel,* and similar to that of other fishes.[7]

---

[1] The trivial name of the first *four* of these species is *electricus.*

[2] *R. Torpedo.*　　　　　[3] *Hist. An. l.* ix. c. 37.

[4] *Hist. Nat. l.* ix. c. 42.　　[5] Heb. רעש

[6] *Silurus.* L.　　　　[7] Geoff. St. Hil. *Ann. du Mus.* i. 402.

This organ is divided into two portions by a longitudinal septum.

The *Torpedo* is the most celebrated of the electric fishes. In this the organ of its power extends, on each side, from the head and gills to the abdomen, in which space it fills all the interior of the body. Each organ is attached to the parts that surround it, by a cellular membrane and by tendinous fibres. Under the skin which covers the upper part of these organs, are two bands, one above the other, the upper one consisting of *longitudinal* fibres, and the lower of *transverse* ones. The latter continues itself in the organ by means of a great number of membranous elongations, which form many-sided vertical bodies, or hollow polygonal tubes, some hexagonal, others pentagonal, and others quadrangular; each of these tubes is divided, internally, by a fine membrane into several dissepiments, connected by blood-vessels. In each of the organs, from two hundred to twelve hundred of these tubes have been counted in individuals of different age and size, some regular but others irregular, which may form electric batteries. Each organ is also traversed by arteries, veins, and nerves, in every direction, which last are remarkable for their size. The tubes, like those above mentioned, are also found in the non-electric Rays, but these terminate in pores without the skin, which are so many excretory organs of the matter contained in their interior; in the Torpedo, on the contrary, the tubes are completely closed, not only by the skin which is no where perforated, but further by the aponeuroses, or tendinous expansions of the muscles, which extend all over the electric organ; the gelatinous matter not being able to expand itself externally, is forced to accumulate in these tubes, from whence doubtless arises their size and their progressive numerical increase. The two surfaces of the electric organ are supposed to be one positive and the other negative. Reaumur observed that the back of the animal is rather convex, but when about to strike its convexity diminishes, and it becomes concave, but after the stroke it resumes its convexity. These organs not only affect the animals upon which they act, by an agency imperceptible to the eye, but they are also stated to emit sparks; and they can strike at some distance, as well as by immediate contact. The author last named put a torpedo and a duck into a vessel filled with sea water, and covered it to prevent the escape of the latter, which, after about three hours, was found dead. These wonderful and complex organs, and their many-phialed batteries, the effect of which has attracted the notice of scientific men for so long a period, were doubtless given to these animals by their Creator, in lieu of the offensive and defensive arms which enable the rest of their tribe to act the part assigned to them, that they might procure the means of subsistence, and to defend themselves when in danger. Almost always concealed in the mud, like most of the *rays*, they can by this weapon kill the small fishes that come within the sphere of their action, or benumb the large ones; if they are in danger of attack from any voracious fish, they can disable him by invisible blows, more to be dreaded than the teeth of the shark itself.

The *Gymnotus*, or electric eel, is a still more tremendous assailant, both of the inhabitants of its own element, and even of large quadrupeds, and of man himself if he puts himself in its way. Its force is said to be *ten* times greater than that of the torpedo. This animal is a native of South America. In the immense plains of the Llanos, in the province of Caraccas, is a city called Calabozo, in the vicinity of which these eels abound in small streams, insomuch that a road formerly much frequented was abandoned on account of them, it being necessary to cross a rivulet in which many mules were annually lost in consequence of their attack. They are also extremely common in every pond from the equator to the 9th degree of north latitude.

Contrary to what takes place in the torpedo, the electric organs of the *Gymnotus* are placed under the *tail*, in a place removed from the vital ones. It has *four* of these organs, two large and two small, which occupy a third of the whole fish : each of the larger organs extends from the abdomen to the tail; they are separated from each other above by the dorsal muscles, in the middle of the body by the natatory vesicle, and below by a particular septum. The small organs lie over the great ones, finishing almost at the same point; they are pyramidal, and separated from the others by membrane. The interior of all these organs presents a great number of horizontal septa, cut at right angles by others nearly vertical. John Hunter counted thirty-four in one of the great organs, and fourteen in one of the small ones, in the same individual. The vertical septa are membranous, and so close to each other that they appear to touch. It is by this vast quadruple apparatus, which sometimes in these animals is calculated to equal one hundred and twenty-three square feet of surface, that they can give such violent shocks. Mr. Nicholson thought that the *Gymnotus* could act as a battery of 1,125 square feet. Humboldt says that its galvanic electricity produces a sensation which might be called *specifically* different from that which the conductor of an electric machine, or the Leyden phial, or the pile of Volta, cause. From placing his two feet on one of these fishes just

taken out of the water, he received a shock
more violent and alarming than he ever experienced from the discharge of a large Leyden
jar; and for the rest of the day he felt an acute
pain in his knees, and almost all his joints.
Such a shock, he thinks, if the animal passed
over the breast and the abdomen, might be
mortal. It is stated that when the animal is
touched with only one hand the shock is very
slight; but when two hands are applied at a
sufficient distance, a shock is sometimes given so
powerful as to affect the arms with a paralysis
for many years. It is said that females, under
the influence of a nervous fever, are not affected.

Humboldt gives a very spirited account of the
manner of taking this animal, which is done by
compelling twenty or thirty wild horses and
mules to take the water. The Indians surround
the basin into which they are driven, armed
with long canes, or harpoons; some mount the
trees whose branches hang over the water, all
endeavouring by their cries and instruments to
keep the horses from escaping : for a long time
the victory seems doubtful, or to incline to the
fishes. The mules, disabled by the frequency
and force of the shocks, disappear under the
water ; and some horses, in spite of the active
vigilance of the Indians, gain the banks, and
overcome by fatigue, and benumbed by the
shocks they have encountered, stretch them-

selves at their length on the ground. There
could not, says Humboldt, be a finer subject for
a painter: groups of Indians surrounding the
basin; the horses, with their hair on end, and
terror and agony in their eyes, endeavouring to
escape the tempest that has overtaken them;
the eels, yellowish and livid, looking like great
aquatic serpents, swimming on the surface of
the water in pursuit of their enemy.

In a few minutes two horses were already
drowned: the eel, more than five feet long,
gliding under the belly of the horse or mule,
made a discharge of its electric battery on the
whole extent, attacking at the same instant the
heart and the viscera. The animals, stupified
by these repeated shocks, fall into a profound
lethargy, and, deprived of all sense, sink under
the water, when the other horses and mules
passing over their bodies, they are soon drowned.
The *Gymnoti* having thus discharged their
accumulation of the electric fluid, now become
harmless, and are no longer dreaded: swimming half out of the water, they flee from the
horses instead of attacking them; and if they
enter it the day after the battle, they are not
molested, for these fishes require repose and
plenty of food to enable them to accumulate
a sufficient supply of their galvanic electricity.
It is probable that they can act at a distance,
and that their electric shock can be communi-

cated through a thick mass of water. Mr. Williams, at Philadelphia, and Mr. Fahlberg, at Stockholm, have both seen them kill from far living fishes which they wished to devour: Lacepede says they can do this at the distance of *sixteen* feet. They are said also to emit sparks.

Of all the *Gymnoti* the *electric* is the only species in which the natatory vesicle extends from the head to the tail; it is in that species of the extraordinary length of two feet five inches, and one inch and two lines wide, but the diameter diminishes greatly towards the tail: it reposes upon the electric organs. It has been asserted that this fish is attracted by the loadstone, and that by contact with it it is deprived of its torporific powers.[1]

It is singular that in the three principal animals which Providence has signalized by this wonderful property, the organs of it should differ so much, both in their number, situation, and other circumstances; but as there appears to be little other connection between them, it was doubtless to accommodate them to the mode of

[1] The authors from whom my information on the electric fishes is chiefly derived are, Rudolphi, *Anatomische Bemerkungen*, &c. 1826; Geoffroy, *Ann. du Mas.* i.; Lacepede, *Hist. des Poissons;* Humboldt, *Observations de Zoologie et d'Anatomie comparée;* and Bosc, in *N. D. D'Hist. Nat.* xii. xiv. xxxiv.

life and general organization of the fishes so privileged.

There is another little fish, of a very different tribe, which emulates the electric ones, in bringing its prey within its reach, by discharging a grosser element at them. It belongs to a genus,[1] the species of which are remarkable for the singularity of their forms, the brilliancy of their colours, and the vivacity of their movements. The species I allude to[2] may be called the *fly-shooter*, from its food being principally flies, and other insects, especially those that frequent aquatic plants and places. These, as Sir C. Bell relates,[3] it, as it were, *shoots* with a drop of water.

In a former part of this treatise I have given an account of those American fishes, which, when the water fails them in the streams they inhabit, by means of a moveable organ, representing the first ray of their pectoral fin,[4] are enabled to travel overland in search of one whose waters are not evaporated. An analogous fact has been observed in China, by a friend and connection of mine,[5] who paid particular attention to every branch of zoology when in the East. At Canton he informed me there is a fish that

---

[1] *Chætodon.*      [2] *C. rostratus.*

[3] *B. T.* 200.      [4] PLATE XII. FIG. 2.

[5] Robert Martin, Esq. F.Z.S.

crosses the paddy fields from one creek to another, often a quarter of a mile asunder. The Chinese told him that this was done by means of a kind of *leg*.

I shall close this history of Fishes with some account of the tribe to which the *fishing-frog*[1] belongs. I have before alluded to their connection with the Reptiles ;[2] in some points also they look to the rays and the sharks. The attenuated tail of all,[3] and the enormous swallow of others,[4] give them this resemblance, especially to the first, so that the French call them *fishing-rays*.[5] The best known of them is that called, by way of eminence, the *fishing-frog*. This is a large fish, sometimes seven feet long ; it is found in all the European seas, and is often called the sea-devil. " This fish," says Lacepede, " having neither defensive arms in its integuments, nor force in its limbs, nor celerity in swimming, is, in spite of its bulk, constrained to have recourse to stratagem to procure its subsistence, and to confine its chase to ambuscades, for which its conformation in other respects adapts it. It plunges itself in the mud, covers itself with sea-weed, conceals itself amongst the stones, and lets no part of it be perceived but the extremity of

[1] *Lophius Piscator.*
[2] See above, p. 389.
[3] PLATE XIII. FIG. 1, 3.
[4] *Ibid.* FIG. 3.
[5] *Raie pécheresse.*

the filaments that fringe its body, which it agitates in different directions, so as to make them appear like worms or other baits. The fishes, attracted by this apparent prey, approach, and are absorbed by a single movement of the fishing-frog, and swallowed by his enormous throat, where they are retained by the innumerable teeth with which it is armed. Another animal of this tribe is furnished only with a single bait, just above the mouth."[1]

We see by this singular contrivance that fertility of expedient by which the Beneficence, and Wisdom, and Power of the Creator have remedied the seeming defects which appear incident to almost every animal form. If it cannot pursue and overtake and seize its prey, it is enabled, as in the case of the *electric fishes*, the *fly-shooter*, and the *fishing-frogs*, in a way we should not expect, to ensure its subsistence ; and while it is doing this, discharging, if I may so speak, its official duty, and acting that part, on its own theatre, by which it best contributes to the general welfare.

Doubtless the infinite forms of the Class we are considering, that inhabit the, so called, element of water, and of which probably we may still be unacquainted with a very large pro-

---

[1] *Malthus Vespertilio*, PLATE XIII. FIG. 1, 2, *a*.

portion, all bear the same relation to each other, and are organized with a view to a similar action upon each other, that we see takes place upon the earth. There are predaceous fishes to keep the aquatic population of every description within due limits ; there are others whose office it is to remove nuisances arising from putrescent substances, whether animal or vegetable ; and lastly, there are others which, like our herds and flocks, are peaceful and gregarious, and graze the herbage of sea-weeds that cover the ocean's bed. All these, in their several stations, and by their several operations, glorify their Almighty Author by fulfilling his will.

# Chapter XXII.

## *Functions and Instincts.   Reptiles.*

In the whole sphere of animals, there are none, that, from the earliest ages, have been more abhorred and abominated, and more repudiated as unclean and hateful creatures, than the majority of the Class we are next to enter upon,—that of *Reptiles*.   One Order[1] of them, indeed, consisting of the turtles and tortoises, and some individuals belonging to another,[2] are exempted from this sentence, and are regarded with more favourable eyes; but the rest either disgust us by their aspect, or terrify us by their supposed or real power of injury.

In Scripture, the *serpent*; the larger *Saurians*, under the names of the *dragon* and *leviathan*; and *frogs* are employed as symbols of the evil spirit, of tyrants and persecutors, and of the false prophets that incite them.[3]

---

[1] The *Chelonians*.

[2] The *Gecko, Monitor, Chamæleon*, &c. amongst the *Saurians*.

[3] *Job*, xli. 34; *Psl.* xxvii. 1; *Ezek.* xxv. 3; *Rev.* xx. 2, xvi. 13.

Yet these animals exhibit several extraordinary characters and qualities. They are endued with a degree of vivaciousness that no others possess: they can endure dismemberments and privations which would expel the vital principle from any creature in existence except themselves. Their life is not so concentrated in the brain, which with them is extremely minute, but seems more expanded over the whole of their nervous system : take out their brain or their heart, and cut off their head, yet they can still move, and the heart will even beat many hours after extraction ; it is also stated that they can live without food for months, and even years.[1]

But though gifted by their Creator with such a tenacity of *life*, yet is that life often raised a very few degrees above death. Many of them select for their retreats damp and gloomy caverns and vaults, shut out from the access of the light and air. In allusion to this circumstance, Babylon, the imperial city, she, who in ancient times subjected the eastern world to her domination, was forewarned that she should *become* heaps, *and a dwelling-place for* dragons.[2]

Whether the many instances that have been recorded in different countries, of *toads* found

---

[1] Cuv. *Règn. An.* ii. 1. 8.　Lacep. *Quad. Ovipar.* i. 20.
[2] *Jerem.* li. 37.

incarcerated alive in blocks of stone or marble, or in trunks of trees, are all to be accounted for by supposing a want of accurate observation of the concomitant circumstances in those that witnessed their discovery, I will not take upon me to say; but they are so numerous, as to leave some doubt upon the mind whether some of these creatures may not have been accidentally interred alive, as it were, when in a torpid state, and continued so, till, their grave being opened, and the air admitted to their lungs again, their vital functions have been resumed, to the astonishment of those who witnessed the seeming miracle. Though so given to withdraw themselves into dark and dismal retreats, yet many of them are fond also of basking in the sun-beam, particularly the serpents and the lizards.

Zoologists seem not even yet fully to have made up their minds with regard to the classification of *Reptiles*. Linné placed them in the same Class[1] with the *Cartilaginous* Fishes, of which they form his first and second *Orders;* but subsequent zoologists, with great propriety, have generally considered them as forming a Class by themselves, under their primeval name of Reptiles. This Class M. Brongniart divided into *four* Orders, viz. *Chelonians, Saurians, Ophidians,* and *Batrachians:* and Baron Cuvier has followed

---

[1] *Amphibia.*

this arrangement in his *Règne Animal*. Latreille, adopting the Group, has divided it into
*two* Classes, *Reptiles* and *Amphibians*. The
Reptiles he considers as forming two *Sub-classes*,
viz. *Cataphracta*, containing the *Chelonians*, and
*Crocodiles*, and *Squamosa*, containing the remaining *Saurians* and the *Ophidians*. His second Class, the *Amphibians*, consisting of the
*Batrachians* of Brongniart, with the addition of
the *Proteus*, *Siren*, &c. he divides into two
Tribes, viz. *Caducibranchia*, or the *proper Batrachians*, and *Perennibranchia*, or the *Proteus*,
*Siren*, *Axolot*, &c. This classification is adopted
by Dr. Grant,[1] except that he does not sub-
divide the Reptiles into two Sub-classes; and
Latreille's two *Tribes* of Amphibians he properly denominates *Orders*.

That Reptiles, in the larger sense of the term,
form a *natural* Group, will be generally admitted,
when it is considered that the *salamanders*, or
naked efts, evidently connect the Batrachians
with the Saurians, and were formerly considered
as a kind of *lizard;* it seems to me therefore
more consistent with nature to consider the
Reptiles as forming a *single* Class.

This opinion has received strong confirmation
from a circumstance communicated to me by
my kind friend Mr. Owen, well known as one of

---

[1] *Outlines of a Course of Lectures, &c.* 14—16.

our most eminent comparative anatomists. In a letter received from him, since I wrote the preceding paragraph, in reply to some queries I had addressed to him, he says,—" I lose no time in replying to your very welcome letter, because I have a statement to make which justifies your disinclination to regard the *Reptilia* of Cuvier as including two distinct Classes. Not any of the *Batrachia* have a *single* auricle ; for though the venous division of the heart has a simple exterior, it is in reality divided internally into two separate auricles, receiving respectively, the one, the carbonized blood of the general system, the other and smaller, the aërated, or vital, blood from the lungs. This I have found to be the case successively in the frog and toad, the salamander and newt, and lastly, in the lowest of the true Amphibia, the *Siren lacertina,* which in its persistent external branchiæ comes nearest, I apprehend, to the Fishes."

By this statement it appears that those characters, which have been deemed sufficient to warrant the division of the Reptiles into two distinct Classes, exist only in appearance. I shall consider them therefore as forming only *one,* of which the following seem to constitute the principal diagnostics.

### Reptilia.  *(Reptiles.)*

*Animal*, vertebrated, oviparous, or ovovivipa-rous.   *Eggs*, hatched without incubation.

*Heart*, really biauriculate, though in some the auricles are not *externally* divided.   *Blood*, red, partially oxygenated, cold.

*Brain*, very small; *vitality*, in some degree, independent of it.

*Integument*, various.

As the two Orders into which the Batra-chians of Cuvier are divided by Dr. Grant, differ from the rest of the Class not only in their respiratory organs, but also in other important particulars, indicating that they form a group of greater value than the other three Cuvierian Orders, I shall therefore consider the Class of Reptiles as further divided into two *Sub-classes*, which I propose to denominate, from the differ-ence of their integument, *Malacoderma* and *Scleroderma*.

*Sub-class* 1.—*Reptilia Malacoderma.* (Soft-coated Reptiles). *Heart*, with two auricles, externally simple, but internally divided.  *Inte-gument*, soft, naked.  *Eggs*, impregnated, *after* extrusion.

This Sub-class consists of the two Orders called, by Latreille and Dr. Grant, as above stated, *Caducibranchia* and *Perennibranchia;* but considering the Reptiles as forming a single

Class, for the sake of concinnity of nomenclature, I think it would be better to restore to the *first* their old name of *Batrachians;* and, as the animals that form the second, as Cuvier observes, are the only true *Amphibians*,[1] to distinguish them by the name that strictly belongs to them alone.

*Sub-class* 2.—*Reptilia Scleroderma.* (Hard-coated Reptiles). *Heart*, with *two* auricles. *Integument*, hard, often scaly. *Eggs*, impregnated *before* extrusion.

### Orders.

| Sub-class 1. | Sub-class 2. |
|---|---|
| 1. *Amphibians.* | 3. *Ophidians.* |
| 2. *Batrachians.* | 4. *Saurians.* |
|  | 5. *Chelonians.* |

*Order* 1.—*Amphibians.* (*Siren, Proteus, Axolot,* &c.)

*Respiration*, double, by gills in the water, and by pulmonary sacs in the air. *Gills*, permanent. Legs, 2—4.

*Order* 2.—*Batrachians.* (*Amphiuma, Triton* or *Water-newt, Salamander, Toad, Frog,* &c.)

*Respiration*, at first by *gills*, and afterwards by *lungs*. *Gills*, temporary. *Ribs*, rudimental. *Legs*, four. Undergoes a *metamorphosis*.

----

[1] *Règne Anim.* ii. 117.

*Order 3.—Ophidians. (Snakes and Serpents.)*

*Body*, covered with scales, without legs. *Ribs*, moveable. *Mouth*, armed with teeth. *Cast* their skin.

*Order 4.—Saurians.* (Two-footed and four-footed *Lizards*, of various kinds; *Crocodiles, Alligators*, &c.)

*Body*, covered with *scales*, or scaly *grains*, terminating in a *tail*. *Ribs*, moveable; *mouth*, armed with teeth. *Legs*, 2—4.

*Order 5.—Chelonians. (Turtles* and *Tortoises.)*

*Body*, protected above by a *carapace*, or shield, formed by the ribs, and below by a *plastron*, or dilated sternum. *Mouth*, without teeth. *Mandibles*, rostriform. *Legs* or *paddles*, four.

Though the *Malacoderm*, or soft-coated Reptiles, appear the legitimate successors of the Fishes, yet there are some others in the higher Orders that seem to lead off towards them also, for the *Ophidians* and *Apod* fishes evidently tend towards each other. The *Cœcilia*, or blind serpent, too, is almost uniauriculate, and has only some transverse rows of scales between the wrinkles of its skin.[1]

From this statement, it seems that the Class of Reptiles is connected with the Fishes, not by those at the top of the latter Class, but by those

---

[1] *Règne Anim.* ii. 99.

at its base ; with the Osseans by the Apods, and with the Cartilagineans by the Cyclostomes ; so that they may be almost regarded as forming a parallel line with them, instead of succeeding them in the same series. Even the proper Batrachians seem to tend to the Chelonians, while the Salamanders look to the Saurians.

The great body of the Class are predaceous, subsisting upon various small animals, especially insects, and some Ophidians upon large ones ; but the *Chelonians* seem principally to derive their nutriment from marine and other vegetables, though some of these will devour Molluscans, worms, and small reptiles : the *Trionyx ferox* will attack and master even aquatic *birds*. Cuvier says, after Catesby, that the common *Iguana* subsists upon fruit, grain, and leaves. Bosc states that it lives principally upon *insects ;* and that it often descends from the trees after earth-worms and small reptiles, which it swallows whole.[1]

*Order* 1.—The *Siren,* or *Mud-iguana,* occupies the first place in this Order, and seems to connect with the Apod and Cyclostomous Fishes, from which it is distinguished by its gills in three tufts, and by having only one pair of legs. It appears to be an animal useful to man, since

[1] *Règn. An.* ii. 44. *N. D. D'H. N.* xvi. 113.

it is stated to frequent marshes, in Carolina, in which rice is cultivated, where it subsists upon earth-worms, insects, and other similar noxious creatures.

But of all the animals which God hath created to work his will, as far as they are known to us, none is more remarkable, both for its situation and many of its characters, than one to which I have before adverted,[1] as affording some proof, that *the waters under the earth*, and other subterranean cavities, may have their peculiar population. The animal I allude to is the *Proteus*, belonging to the present Order, which was first found thrown up by subterranean waters in Carniola, as we are informed by the late Sir H. Davy,[2] by Baron Zöis. Sir Humphry himself appears to have found them in the Grotto of the Maddalena, at Adelsburg, several hundred feet below the surface of the earth; he also states that they have been found at Sittich, thirty miles distant, and he supposes that those found in both places might be thrown up by the same subterranean lake.[3] In the year 1833 there were two living specimens in the museum of the Zoological Society, where I had the pleasure of seeing them; and from one of them the accurate figure at the end of this volume,[4] by

[1] VOL. I. p. 35.  
[2] *Consolat. in Trav.* 187.  
[3] *Ibid.* 183—188.  
[4] PLATE XIV. FIG. 1.

the kind permission of the Society, was taken by Mr. C. M. Curtis.

When we look at these animals, there is something so different in their general aspect from the tribes to which they are most nearly related, that the idea strikes one that we are viewing beings far removed from those that inhabit the surface of our globe, and its waters; which, though accidentally visiting these upper regions, may be the outsetters of a population still further removed from our notice, and dipping deeper into its interior.

The *Proteus* is about a foot in length, or something more, and about an inch in thickness; the body is cylindrical, tapering to the tail; its colour is a pale red; its skin is transparent and slimy, so as easily to elude the grasp. It has four short slender legs, the anterior pair placed just behind the head, having *three*, and the posterior pair, which are shorter, and placed just before the vent, having only *two* toes without claws. The head terminates in a flat, very obtuse muzzle, somewhat resembling the beak of a duck; its maxillæ are armed with teeth; the eyes are extremely minute, and scarcely discernible; they are concealed, and apparently rendered useless by an opaque skin; but as this animal is said to avoid the light, it is evident that it produces some effect upon them; behind the head, on each side, is an opening like those of fishes, over which are the gills, divided into several branches.[1] It has, besides, an internal pneumatic apparatus, consisting of two vesicles, below the heart. The tail is compressed, furnished above and below with a caudal fin, extending to the posterior legs. Its legs, from their having no claws, are, it is probable, principally useful in walking upon the mud, and by means of its caudal fin it can move like an eel or fish in the water. From a small shell-fish being found in the stomach of one, it seems to follow that its food, at least in part, consists of Molluscans inhabiting the same subterranean caves and waters with itself, and probably distinct from any of those to which the atmosphere has free access. Sometimes, elevating its head above the water, it makes a hissing noise louder than could be expected from so small an animal.

Before quitting this subject, I may observe that Baron Humboldt has given an account of a wonderful eruption of *subterranean fishes*, which sometimes takes place from the volcanos of the kingdom of Quito. These fishes are ejected in the intervals of the igneous eruptions, in such quantities as to occasion putrid fevers by the miasmata they produce: they sometimes issued from the crater of the volcano, and sometimes

[1] PLATE XIV. FIG. 1, *a.*

from lateral clefts, but constantly at the elevation of between two and three thousand toises above the level of the sea. In a few hours, millions are seen to descend from Cotopaxi, with great masses of cold and fresh water. As they do not appear to be disfigured or mutilated, they cannot be exposed to the action of great heat. Humboldt thought they were identical with fishes that were found in the rivulets at the foot of the volcanos. These fishes belong to a genus separated from *Silurus*.[1]

*Order* 2.—This Order begins with two genera, the species of which have been supposed to breathe by lungs only, no traces of gills having yet been discovered in any individual belonging to them. Cuvier thinks that they cast them sooner than the salamanders. One of these is a large animal,[2] being more than a yard in length; it was discovered by Dr. Garden, in South Carolina: like the Proteus, its eyes are covered with a thick tunic, and its toes have no claws. The other,[3] found in New York, comes near the salamanders, and has been called by American writers the *giant* salamander. Both are found in fresh-water lakes, and similar places.

I have mentioned, on a former occasion, a salamander that lays her eggs singly on the

---

[1] *Pimelodus.* Humboldt names the species in question *P. Cyclopum. Zool.* 22.

[2] *Amphiuma means.*     [3] *Menopoma.*

leaves of *Persicaria*, which she doubles down over them,[1] and which are kept folded by means of the glue that envelopes the egg. Dr. Rusconi, to whom we are indebted for this history, observed the whole progress and developement of this animal, from its embryo state in the egg. It is at first opaque, formed of a soft homogeneous substance. Almost as soon as it has escaped from its envelope, it becomes gradually transparent, so that the successive developements, both of its internal and external organs, may be discerned—the heart, and its systole and diastole; the stomach, its form and position; the intestinal canal, which at first extends in a straight line, from one end of the abdomen to the other, and then begins to undulate, and ends by forming many convolutions: next may be seen the liver, the developement of which keeps pace with that of the stomach and intestines; and lastly appear the lungs, taking their place and form, always filled with air, and so transparent that one might believe the animal has on each side of the trunk a bubble of air gradually dilating and lengthening. When all these organs have acquired the necessary developement, the spectator beholds in the little creature the beginning, as it were, of its animal

---

[1] See above, p. 265.

life. Its former life being merely organic, resembling that of a vegetable, but now its motions are become the result of its sensations and will.[1]

We see in this instance how exactly the rudiments, as it were, of the organs of the future animal, are fitted to respond to the action of the elements upon them, how the germe of every organ begins, if I may so speak, to vegetate, and grows till it is fully developed, so as to become either a fit instrument of the will or of the vital powers, and adapted to carry the creature through all its destined operations, and to enable and incline it to fulfil all its prescribed functions. These observations, and this interesting little history, will apply to *man* himself, who, in his embryo state, is the subject of similar developements; and the words of the divine Psalmist are a beautiful comment upon this our embryo life : *For thou hast possessed my reins : thou hast covered me in my mother's womb. My substance was not hid from thee, when I was made in secret, and curiously wrought in the lowest parts of the earth. Thine eyes did see my substance yet being imperfect ; and in thy book all my members were written, which in continuance*

---

[1] Rusconi, in *Edinb. Philos. Journ.* ix. 110—113, on *Salamandra platycauda.*

*were fashioned, when as yet there was none of them.*[1]

The salamander, as is reported, says Aristotle, if it goes through fire extinguishes it :[2] this is repeated by Pliny, who adds, that it extinguishes it like ice. It never appears, he further observes, except in showery weather, and likewise that it emits a milky saliva, which is depilatory.[3] Salamanders, says Bosc, emit from their skin a lubricating white fluid when they are annoyed, and if they are put into the fire, it sometimes happens that this fluid extinguishes it sufficiently to permit their escape; and again—when one touches the terrestrial salamander, it causes to transude from its skin a white fluid, which it secretes more copiously than its congeners. This kind of milk is extremely acrid, and produces a very painful sensation upon the tongue. According to Gesner, it is an excellent depilatory. It is sometimes spirted out to the distance of several inches, as Latreille has observed, and diffuses a particularly nauseous scent; it poisons small animals, but does not appear to produce serious effects upon large ones.[4]

I have introduced these ancient and modern statements to show how little they differ, and in confirmation of the truth of them I have a re-

---

[1] *Ps.* cxxxix. 13—16.　　[2] *Hist. An.* lib. v. chap. 19.
[3] *Hist. Nat.* l. x. 67.　　[4] *N. D. D'H. N.* xxx. 58, 59.

markable occurrence to relate, which I give
upon the authority of three ladies who witnessed
the fact, and upon whose accuracy I can rely.
They were residing at Newbury, where their
cellars were frequented by frogs, and a kind of
newt, or salamander, of a dull black colour.
Several of the frogs were caught one day, and
put into a pail; and while the ladies were
looking at them they were surprised by observing the frogs one after another turn themselves on their backs, and lie with their legs
extended quite stiff and dead. Upon examining the pail they found one of these efts, as they
called them, running round very quickly amongst
the frogs, each of which, when touched by it,
died instantaneously, in the manner above
stated. They afterwards regarded these efts,
as may be supposed, with nearly as much
horror as they would a rattlesnake; and a few
nights afterwards, finding one in the kitchen,
it was seized with the tongs, and thrown into a
good fire which was burning in the grate. The
reptile, instead of perishing, slipped like lightning through the coals, and ran away under the
fireplace apparently unhurt. The house, in
which these animals were found, was in a remarkably damp situation.

If our northern salamanders are gifted with
such powerful means of offence or defence, we
know not how far those powers may be sublimed

in the species of warmer climates ; and the fire-quenching and death-doing properties of the Grecian or Roman salamanders may approach nearer to the, supposed, fabulous descriptions of Aristotle and Pliny, than modern Herpetologists seem willing to believe.

There appears no small analogy between these properties considered as weapons, and means by which these animals either secure their prey, consisting of earth-worms, insects, and other small game, or disarm and destroy their enemies, and those, related in the last chapter, which distinguish the electric fishes.

Spallanzani, by numerous experiments, has discovered in this tribe of animals, the power of reproducing lost or mutilated organs ; Bonnet and others have confirmed his observations. So that it seems proved, if their legs and tail are cut off, and even their eyes plucked out, that in a few months they will be reproduced ; and even a limb thus renewed, if again cut off, will be reproduced again.

In going upwards from the salamanders, at first sight, we feel disposed to proceed next to the other animals of a similar form, the *lizards* and other Saurians, for this way their external form leads us, but their internal organization is nearer that of the frogs and toads.   Upon these last I shall not dwell : all know that they begin life in the water like fishes ; that they are at first without legs, or any instrument of motion but a tail, which by its undulations from side to side steers the apparently disproportioned body to which it is appended, and makes its way with rapidity through its native element. Few are ignorant that they first acquire a single pair of legs; and lastly, that, another pair being also acquired, they leave the water by myriads, and appear, without a tail, as four-footed, and, at certain times, noisy reptiles.

*Order* 3.—The general function of the *Ophidians* seems connected with almost the whole animal kingdom. The insects, frogs, and other reptiles, several birds and beasts, up as high as the *ruminant* and even the *carnivorous* tribes, become the prey of various species. They act the same part with land animals, that their analogues, the eels and other apod and cyclostomous fishes do with respect to those of the water. Some are analogues of the lion and the tiger, as the *Oriental Python* and the *Occidental Boa*, which sometimes exceed thirty feet in length, and are as thick as a man's body; while others compete with the minor predaceous beasts in the destruction they occasion amongst the lesser quadrupeds. But while the predaceous quadrupeds, with the exception of the *Hyena*, leave untouched the skeleton of the animals they devour, the Ophidians swallow the entire animal, flesh and bone and skin, and thus completely remove it from the face of nature ; whereas the others, where they abound and are unmolested, make their domain like a charnel house, and deform the earth with the ghastly relics of their cruelty and voracity.

The mechanism of the mouth of these animals is so contrived by Divine Wisdom, and the pieces that form it so put together, as to enable them to twist and distort and dilate it so enormously that they can swallow animals bigger than their own bodies.[1] The vertebræ of the great Boa are more numerous than those of other serpents, which gives them a greater power of surrounding and strangling their prey with their dreadful voluminous folds, of crushing it, and, with the help of their saliva, rendering it fit for deglutition. With their tail, likewise, they can lay strong hold of a tree, so as to use it as a fulcrum, by which their powers of compression are increased and rendered more available where they have to contend with the struggles of powerful animals.

*Order* 4.—The connection of the *Saurians,* or the animals forming the next Order with the Ophidians, is very intimate. Cuvier says that many serpents under the skin have the vestige of a posterior limb, which in some *shows* its extre-

---

[1] Cuv. *Anat. Comp.* iii. 90.

mity externally, in the form of a little claw.[1]
Amongst the lizards is one that has only two
*fore legs*,[2] and another that has only two *hind*
ones ;[3] and a third,[4] in which the legs are so
short and so distant, and the body so slender
and serpentiform, that they resemble a snake
with four legs rather than a lizard.

This Order is divided into numerous genera
and sub-genera. One of the most celebrated is
the Chameleon. I have already noticed some
of its peculiarities, and its mode of catching the
insects that form its food.[5] The ancients were
of opinion that it lived upon air, led by the
power it has of swelling itself to twice its natural
size, by inflating its vast lungs, when its body
becomes transparent. Cuvier is of opinion that
it is the size of the lungs of these animals that
enables them to change their colour, not in order
to assume that of the bodies on which they
happen to be, but to express their wants and
passions. He supposes that the blood, being
constrained to approach the skin, more or less,
assumes different shades, according to the degree
of transparency.[6] The Rev. L. Guilding, however, mentions another genus,[7] the species of
which, when in search of prey, adapt their
colour to the green tree or dark brown rock on

---

[1] *Règn. An.* ii. 71.     [2] *Chirotes.*     [3] *Bipes.*
[4] *Seps.* See Roget, *B. T.* i. 448. *f.* 210.
[5] See above, p. 192.     [6] *Règn. An.* ii. 59.     [7] *Anolis.*

which they lie in ambush.[1]  As these animals have the power of inflation, at least partially, by assuming a degree of transparency, they may appear of the colour of the substance they are standing upon, a remark which may also apply to the chameleon.  The object of this may be to conceal themselves from their enemies, as well as from their prey.

The *Guanas*,[2] also, are said to change their colour ; they are remarkable, as well as the *Anolis*, for the kind of goitre in their throat, which when irritated or excited they can inflate to a large size.  These animals, though their flesh is said to be unwholesome, in the countries they frequent are highly prized for the table, and are often hunted with dogs.  Their eggs also are in request.

The *Monitors*, or *safeguards*, as the French call some of them, deserve notice, because one species[3] is said to assist in the diminution of the crocodile, since, like the ichneumon, it devours its eggs, and even the young ones, on which account it is supposed to be sculptured on the monuments of the ancient Egyptians.  This name was given them because they were believed to warn people, by hissing, of the approach of the crocodile, or venomous reptiles.

---

[1] *Zool. Journ.* iv. 165.  [2] *Iguana vulgaris.*
[3] *M. niloticus.*

But the most celebrated of the Saurians, from the earliest ages, is the *Crocodile:* its history, however, is so well known that I shall only mention a few circumstances, of less notoriety, connected with it. There has been some difference of opinion as to whether the crocodile can move the upper or lower jaw. Aristotle observes, all animals move the lower jaw, except the crocodile of the river, for this animal only moves the upper.[1] Denon says the same.[2] Lacepede, on the contrary, affirms that the lower jaw is the only moveable one.[3] I was assured by Mr. Cross, when looking at two alligators in his menagerie, then at Charing-cross, that they moved both their jaws; and my friend Mr. Martin has observed the same thing in India. M. Geoffroy St. Hilaire and Baron Cuvier nearly reconcile the two opinions. The *head,* says the former, moves on the lower jaw like the lid of a snuff-box, that opens by a hinge. By this mechanism they can elevate their nostrils above the water, which they do with great rapidity for concealment:[4] and the latter observes, that the upper jaw moves only with the whole head.[5] So that the fact seems to be that the lower jaw alone has motion independent of the head, and the upper one can only move with it: but when we

[1] *Hist. An.* lib. i. c. 11.  
[2] *Voyage,* &c. i. 185.  
[3] *Hist. Ov.* 194.  
[4] *An. du Mus.* x. 376.  
[5] *Règn. An.* ii. 18.

consider that the lower one extends beyond the skull, a condyle of which acts in an acetabulum of that jaw, we can easily comprehend that the upper jaw and head forming one piece, may be elevated at any angle, according to the will of the animal; and thus the upper one acquires additional power of action in attacking its prey in the water and securing it.

The nostrils of this animal are at the end of the muzzle, and this structure enables it, by causing the upper jaw to emerge a little, which, as the crocodile cannot remain under water more than ten minutes, enables it to breathe without exposing itself to observation. When on shore it turns itself to the point from which the wind blows, keeping its mouth open. Adanson relates that he once saw in the Senegal more than two hundred of these river monsters swimming together, with their heads only emerging, and resembling so many trees. Were it not for the number of their enemies, great and small, their increase would be so rapid that they would drive man from the vicinity of the great rivers of the torrid zone. The River-horse [1] attacks them and destroys many—Behemoth against Leviathan,—for though the Leviathan of the Psalmist is clearly a marine animal or monster,[2] that of Job [3] is as clearly the croco-

---

[1] *Hippopotamus.*    [2] *Psl.* civ. 26.    [3] Chap. xli.

dile,[1] and they are stated to destroy many of them; even the feline race, in some countries, contrive to make them their prey. Though the scales that cover their back are impervious to a musket ball, those on the belly are softer and more easily penetrated; and here the saw-fish, and other voracious fishes, find them vulnerable, and so destroy them. The *Trionyx*, also, a kind of tortoise, devours them as soon as hatched. Their eggs are the prey not only of the ichneumon and the lizard, before mentioned, but of many kinds of apes; and aquatic birds also devour them, as well as man himself.

The crocodile has no lips, so that when he walks or swims with great calmness, he shows his teeth as if he was in a rage. When extreme hunger presses him, he will swallow stones and pieces of wood to keep his stomach distended. The heron and the pelican are said to take advantage of the terror which the sight of the crocodile produces amongst the fishes—causing them to flee on all sides—to seize and devour them: therefore they are frequently seen in his vicinity.

*Order 5.*—The *Chelonians*, as far as at present known, seem far removed from the Saurians. The turtles, indeed, in their paddles, exhibit an organ which is common to them,

---

[1] VOL. i. p. 30.

and some of the fossil Saurians, as the *Icthyo-saurus* and *Plesiosaurus*. Cuvier places the *Trionyx* next above the crocodiles; but it agrees with them only in its fierceness and voracity, and the number of its claws.

The importance of the highest tribe of this Order to seamen in long voyages, is universally known and acknowledged, but otherwise there is nothing particularly interesting in their. history, or that of the tortoises.

A singular circumstance distinguishes the animals of this Class,—very few of them have teeth formed for mastication. The *guana* is almost the only one amongst the existing tribes that has them. The Chelonians, which seem almost capable of living without food, have none. The teeth of the predaceous tribes are fitted to retain or lacerate their prey, but not to masticate it; so that the function of the great majority appears to be the same with that of the Ophidians before mentioned, the complete deglutition of the animals their instinct compels them to devour. Insects, which, of all minor animals, are the most numerous, and require most to be kept in check, form the principal part of the food of a large proportion of them. Creatures also that frequent dark and damp places, and that take shelter under stones and similar substances, seem to be particularly appropriated to them by the will of their Creator. Of this description are slugs, earth-worms, and several others : these, therefore, they have in charge to keep within due limits. And thus, in their doleful retreats and hiding-places, they fulfil each its individual function, instrumental to the general welfare.

## CHAPTER XXIII.

### *Functions and Instincts.   Birds.*

WE are now arrived at the highest department of the animal kingdom, the members of which are not only distinguished by a vertebral column, but also by *warm* red blood, and a more ample brain. This department consists of two great Classes, viz. those that are *oviparous*, and do *not* suckle their young; and those that are *viviparous*, which suckle their young till they are able to provide for themselves. The first of these Classes consists of the *Birds*, and the last of the *Quadrupeds*, *Whales*, and *Seals*, called from the above circumstance *Mammalians*. Man, though *physically* belonging to the latter Class, *metaphysically* considered, is placed far above the whole animal kingdom, by being made *in the image and after the likeness* of his Creator, receiving from him *immediately* a reasonable and immortal soul; and entrusted by him with *do-minion over the fish of the sea, and over the fowl of the air, and over every living thing that moveth upon the earth.*

Having, in a former chapter, given some account of those animals, to which the *waters* of this globe are assigned as their habitation and scene of action, I am now to consider those which their Creator has endowed with a power denied to man, and most of the Mammalians— that of moving to and fro in the air as the fishes do in the water, which, on that account, though they move also on the earth, are denominated, in the passage just quoted, the fowl of the *air.*

The animals of this great Class are rendered particularly interesting to man, not only because many of them form a portion of his domestic wealth, look to him as their master, and vary most agreeably his food; but because numbers, also, strike his senses by the eminent beauty and grace of their forms, the brilliancy or variety of the colours of their plumage, and the infinite diversity, according to their kinds, of their motions and modes of flight. But of all their endowments, none is more striking, and ministers more to his pleasure and delight, than their varied song. When *the time of the singing* birds *is come, and the voice of the turtle is heard in our land,* who can be dead to the goodness which has provided for *all* such an unbought orchestra, tuning the soul not only to joy, but to mutual goodwill; reviving all the best and kindliest feelings of our nature, and calming, at least for a time, those that harmonize less with the scene before us.

I may here offer a few observations upon the *voice* of animals, especially birds. A distinction is made by physiologists between a *voice* and a *sound*, and none but those that breathe by means of *lungs* are reckoned to utter a *voice;* others, whatever their respiratory organs, only emit a *sound.* The voice also is from the *mouth* alone, the sound from *other* parts of the body.[1] The *vocal* animals, therefore, are confined to the three last classes of vertebrates—the *Reptiles,* the *Birds,* and the *Mammalians.* In most of these, also, the voice partakes, in some degree, of the character of *speech ;* it is intended to indicate to another the wishes, emotions, or sufferings of the utterer. The great organ of the voice is the *wind-pipe,* or tracheal artery, as it is often called, and its parts, which by its bronchial ramifications is so intimately connected with the lungs as to form part of their substance.

*Birds,* of all animals, are best organized with regard to their voice. Besides the upper larynx, or throat, which they have in common with Mammalians, at the base of their windpipe, where it divides into two branches, render-

[1] See *Introd. to Ent.* Lett. xxiv.

ing to each lobe of the lungs, it has also another larynx, forming a second vocal apparatus. This is produced by a contraction of the organ furnished with muscular fibres, or vocal strings, which by their various tensions and relaxations, modify greatly the tones of the voice; ascending also in the tube of the wind-pipe to undergo another modification at the upper larynx, which, as it were, adds the tube of the *horn* to that of the *reed*. Thus, if the head of a duck is cut off, it can produce sounds by means of its lower throat, if I may so call it, which no quadruped could do. Besides this, birds can, more or less, shorten or lengthen the tube of their wind-pipe, so as to modify the sounds they emit.

Though the upper larynx, in birds, has no vibratory vocal strings, as in the Mammalians, to modify the sounds, these modifications taking place at the lower larynx, still they can enlarge or contract it, which may affect the air in its exit, and so produce some diversity.

Besides all this, whoever casts an eye over Dr. Latham's and Mr. Yarrel's figures of the wind-pipes of various birds,[1] especially wild-fowl, will see that they vary greatly in their relative length and volume; that some are partially dilated, and others contracted, with other peculiarities that distinguish individual species, espe-

---

[1] *Linn. Trans.* iv. *t.* ix.—xv.; xv. *t.* ix.—xv.; and xvi. *t.* xvii.—xxi.

cially in male birds. All these, no doubt, modify
the voice, and, by the will of Him who formed
them, cause them to utter such sounds, and speak
such a language, as are required by the circumstances in which they are placed. The cawing
of the rook, the croaking of the raven, the cooing
of the dove, the warbling of the nightingale and
the other singing birds, are all the result of their
organization according to the plan and will of
that Supreme Intelligence, infinite Love, Wisdom, and Power, which fabricated and fashioned
them with this view as well as others, to give
utterance to sounds that, mixed or contrasted,
would produce a kind of universal concert, delighting the ear by its very discords.

It is said by a late writer, that the song of the
same individual species of birds, in different districts, is differently modified. This, I should
think, must be occasioned by a difference in the
temperature, and other circumstances connected
with the atmosphere.

Of all animals, birds are most penetrated
by the element in which they move. Their
whole organization is filled with air, as the
sponge with water. Their lungs, their bones,
their cellular tissue, their feathers—in a word,
almost every individual part, admit it into their
interstices.[1] Thus giving them a degree of spe-

[1] *N. D. D'Hist. Nat.* xxiii. 352.

cific levity that no other class of animals is
endowed with, which however does not render
them the sport of every wind that blows, for, by
means of their vigorous wings, formed to take
strong hold of the air; of their muscular force,
the agility of their movements, and their powers
of steerage by means of the prow and rudder of
their little vessel, their head and tail, they can
counteract this levity; and by these also, and by
their great buoyancy, they can ascend above the
very clouds, as well as descend to the earth;
they can glide motionless through the air, or
skim the surface of the waters; they can sport,
at will, in the vast atmospheric ocean; they can
dart forward in a straight line, or like the butterfly, fly in a zigzag or undulatory one, and with
ease take any new direction in their flight that
fear or desire may dictate. Enveloped in soft
and warm plumage, they can face the cold of
the highest regions of the air; and the denser
clad aquatic birds can also sail over the bosom
of the waters, or plunge into them, without being
wetted by them. All birds, especially those last
mentioned, have a gland secreting an oily fluid,
with which they anoint their feathers and repel
the moisture.

There is no part of the history of these animals, in which the care of a fatherly Providence
is more signally conspicuous than their love of

their young, and their tender care of them till they can shift for themselves. But as I have already adverted to this subject,[1] and shall hereafter have occasion to resume it, I shall now say something on the *classification* of the feathered race. It is singular that two Classes should be placed in apposition to each other, seemingly so opposite in their character and most of their qualities, as the Reptiles and the Birds—the one the most torpid and doleful and hateful of animals, symbols of evil demons; the other the most lively and active, and beloved of all the creatures that God has made, symbols of the angelic host, and calling upon us to look upwards, and seek those joys that are above us. But in spite of this apparently striking contrast, still there is a real affinity between the Birds and the Reptiles; and when we recollect that demons are fallen angels, we may apprehend why God has placed their symbols in the same series.

Zoologists are not altogether agreed as to which of the Reptiles come the nearest to the Birds: the beak, and some other characters of the *Chelonians*, have been thought to indicate that they are entitled to that distinction;[2] and, by his placing the latter immediately after the Birds, this appears to be Baron Cuvier's opinion. Any one, indeed, that looks either at the common,[3] or

[1] See above, p. 261—264.
[2] Mac Leay. *Hor. Entomol.* 263.     [3] *T. Mydas.*

the hawk's bill, turtles,[1] or a good figure of them,[2]
will see in them a striking resemblance of some
sea-bird, especially a penguin ; the anterior elongated paddles imitating the *wings*, and the posterior dilated ones the *webbed feet* of such birds.
There are other Reptiles, however, that dispute
this claim with the Chelonians. Amongst the
rest is a remarkable fossil genus, regarded as
extinct, which Cuvier has arranged with the
*dragon* of modern Herpetologists, under the
name of *Pterodactyle*.[3] The carpal and metacarpal bones, and the phalanges of the fourth
toe of the anterior leg are excessively elongated,
to which it is conjectured a membrane was attached, forming a wing for flight.    M. Sömmer-
ing classes this remarkable animal with the
*Mammalians*, supposing its affinity to be with the
*Cheiropterans*, or Bats ; and Dr. Wagler considers it as forming, with the *Echidna* and *Ornithorhynchus*, an osculant Class, which he distinguishes by the ancient name of *Griffins*.[4]
But the wing in its structure appears to approach
nearer to that of *birds*, and therefore Blainville
seems right in considering it as a Saurian genus
leading to them.[5]    Professor Goldfuss, in his

---

[1] *T. Caretta.*        [2] *N. D. D'H.N.* xxxiv. *t.* R. 8. *f.* 1. 2.
[3] *Pterodactylus.    Ornithocephalus.*   Sömm.
[4] *Gryphi.*   Gray's *Synops. Rept.* 78.
[5] *N. D. D'H. N.* xxviii. 226.

description of a new species,[1] mentions having found upon it some impressions, looking like those of feathers; and though he thinks it flies like a bird, seems to regard it as between the crocodile and the monitor. The serrated beak of the mergansers is not very unlike that of the common pterodactyle,[2] though that of the species described by Professor Goldfuss has a few very long dispersed teeth, of different lengths, like those of the crocodile.[3] The animals of the last named genus, in the structure of their heart, approximate most nearly to birds, and in their general organization are at the head of the Class of Reptiles.[4]

From these statements, it seems as if the Class just mentioned sent forth several branches towards the Birds; but, all circumstances considered, the pterodactyle, especially if it has feathers, or rather plumiform scales, appears to come the nearest to them, and to prove that the *feathers* of the *Bird* are a transition from the *scales* of the *Reptile*.

---

[1] *Pt. crassirostris. Isis* Heft. v. 553.

[2] *Pt. antiquus.*

[3] Isis. *ubi supr. t.* vi. *f.* vii.

[4] For these observations, with respect to the crocodile, I am indebted to Mr. Owen.

## AVES. *(Birds.)*

*Animal*, vertebrated, oviparous, biped.
*Anterior extremities*, organized for flight.
*Integument*, plumose.
*Eggs*, usually hatched by incubation.
*Lungs*, fixed.
*Respiration* and *circulation*, double.
*Blood*, red, warm.

Ornithologists appear at present undecided as to the division of this great and interesting Class into *Orders*, as the following synoptical table of systems, differing in this respect, will show:

| | |
|---|---|
| Nitzsch and Schoepss have only ........ | 3 |
| Vieillot, Vigors, Mac Leay and Swainson.. | 5 |
| Linné, Cuvier, Dumeril and Carus........ | 6 |
| Illiger ............................... | 7 |
| Scopoli, Latham, Myers and Wolf......... | 9 |
| Temminck .......................... | 13 |
| Grant .............................. | 16 |
| Schœffer............................ | 17 |
| Brisson ............................ | 28 |
| Lacepede ........................... | 38 |

Orders.

One may truly say here, "the choice perplexes;" and the young Ornithologist must be puzzled to determine which of these systems he ought to adopt, especially since the several authors of them were amongst the most eminent zoologists of their time.

I am indebted to Mr. Owen for my knowledge of the first of these systems, of which, as at present it is little known in this country, I will here give an abstract, without entering into its merits, except that its *primary* sections, or Orders, form a very natural division of the Class.

Orders.—I. Aërial Birds. *Luftvögeln.*
              Sub-orders.—A. Accipitrines.
                       B. Passerines.
                       C. Pies.
   —   II. Terrestrial Birds. *Erdvögeln.*
                       A. Columbines.
                       B. Gallinaceans.
                       C. Coursers.
   —   III. Aquatic Birds. *Wasservögeln.*
                       A. Waders.
                       B. Anserines.

In this last Order he includes the Bustards,[1] which surely ought to form a separate Sub-order.

On the present occasion I shall follow the system of Linné, as improved by Baron Cuvier, in the last edition of his *Règne Animal*, adopting from Illiger his Order of *Cursores*, or runners, which appears to be osculant between the *gallinaceous* Order and that of the *waders*.

That the series ought to begin with the *web-footed* Birds, as approaching nearest to the Reptiles, there is no doubt; but which should terminate it, seems not satisfactorily determined.

[1] *Otis.*

The *birds* of *prey* appear naturally to connect with the *beasts* of *prey*, rather than with the Cetaceans, next before which Cuvier has placed them; Carus ends the series with the Gallinaceans, which Linné *contrasts* with the Ruminants, and Mr. W. S. Mac Leay *connects* with the Gnawers,[1] and Illiger and Lacepede end with the Psittaceans, which are analogues of the Quadrumanes, but these are probably mostly analogous forms; there seems a more strict affinity between the web-footed birds and the *Monotremes*, the *Ornithorhynchus*, *Echidna*, &c. which, in some respects, appear to form an osculant *Order*, between the birds and the beasts. In fact the Birds, though united into one group with the Beasts by common characters, may be regarded as forming a *parallel* series with the latter rather than a *continuous* one, several of the members of which, respectively, represent each other, both as to many of their external features, and their functions. Branches, like those of a tree, seem indeed to issue from every natural series, whether vegetable or animal, on all sides, and to run in all directions towards those of other series, so as to form together a perplexing labyrinth, to thread which, although in many places there appears an evident clue, in others it becomes evanescent, and the investigator of nature seems lost. But when we reflect that the Author of

---

[1] *Rodentia.*

Nature is *infinite* in his essence and attributes, we must expect there will be something that indicates their origin from such a Being; though not a real, there will be in them a seeming infinity to finite minds. He who made them sees them all at once, and in their several places, and traces simultaneously every series through all its numberless divarications or convolutions; whereas man sees only a *part* of the ways of his Creator. He can have no simultaneous view of things, and must be contented with adding, here a little and there a little, to his stores of knowledge. To investigate the works of his Creator is a laudable exercise of his powers, and to aim as much as possible to discover the system of things that the God of Nature has established by his Wisdom, and upholds by his Power, is to aim at the discovery of Truth; who will more and more reveal herself to those that, using the proper means, seek her in sincerity.

### Orders.[1]

| | |
|---|---|
| 1. *Swimmers.* | 5. *Climbers.* |
| 2. *Waders.* | 6. *Perchers.* |
| 3. *Coursers.* | 7. *Raveners.* |
| 4. *Scratchers.* | |

[1] The Latin names of the Orders are,—

| | |
|---|---|
| 1. *Natatores.* | 5. *Scansores.* |
| 2. *Grallatores.* | 6. *Insessores.* |
| 3. *Cursores.* | 7. *Raptores.** |
| 4. *Rasores.* | |

\* *Raptor milvius.* Phædr.

*Order* 1.—*Swimmers.* (*Web-footed,* or *Aquatic Birds.* This Order includes the *Inertes, Palmipedes,* and *Pinnatipedes* of Dr. Grant's catalogue.)

BODY, closely covered with feathers, and coated with a thick down next the skin. *Legs,* placed behind the equilibrium. *Toes,* united by membrane for swimming; *membrane* sometimes divided.

*Order* 2.—*Waders.* (*Flamingo, Coot, Avocet, Woodcock, Snipe, Ibis, Spoonbill, Jabiru, Bittern, Heron, Crane, Stork, Oyster-catcher, Plover, Bustard.*—*Grallatores.* Grant.)

*Legs* consisting of very long *tarsi,* with the apex of the *tibia* bare; stretched out in flight. *Wings,* long.

*Order* 3.—*Coursers.* (*Apteryx, Ostrich, Emeu, .Cassowary, Dodo,* &c.—*Cursores.* Grant.)

WINGS, very short, not used for flying. *Legs,* robust. *Toes,* 3—4. *Beak,* depressed or compressed.

*Order* 4.—*Scratchers.* (*Pigeon, Quail, Partridge, Common Poultry, Guinea-fowl, Pheasant, Turkey, Peacock,* &c.—*Alectorides, Gallinæ,* and *Columbæ.* Grant.)

*Upper mandible,* vaulted; *nostrils,* pierced in a membranous space at their base, covered by a cartilaginous scale. *Tail-feathers,* 14—18.

*Order* 5.—*Climbers.* (*Psittaceans, Toucan,*

*Cuckoo, Wryneck, Woodpecker,* &c.—*Chelidones, Alcyones, Anisodactyli, Zygodactyli.* Grant.)

*Feet* with two toes before and two behind.

*Order* 6. — *Perchers.* (*King-fisher, Hoopoe, Humming-bird, Tree-creeper, Nut-hatch, Bird of Paradise, Crow, Magpie, Starling, Cross-beak, Gross-beak, Gold-finch, Linnet, Sparrow, Titmouse, Lark, Goat-sucker, Swallow, Taylor-bird, Nightingale, Red-breast, Fly-catcher, Black-bird, Chatterer, Butcher-bird,* &c.—*Granivoræ, Insectivoræ,* and *Omnivoræ.* Grant.)

*Toes* four: formed for prehension in nidification. *External toe* united at the base to the *internal. Three* toes before and *one* behind. All other characters negative.

*Order* 7. — *Raveners.* (*Owl, Secretary-bird, Buzzard, Kite, Sparrow-hawk, Falcon, Harpy, Eagle, Vulture,* &c.—*Rapaces.* Grant.)

*Beak* robust, upper mandible, on each side, armed with a tooth. *Legs* short, robust. *Toes* armed with crooked claws.

*Order* 1.—The *swimmers,* or web-footed birds, form a very important part of the feathered race, both as furnishing man with food, and as ministering greatly to his comfort, by their down and feathers, when he retires to rest; and also by their action upon the inhabitants of the waters both of the sea and rivers, which form the principal part of their food. Cuvier remarks, that these are the only birds in which the neck exceeds, and sometimes considerably, the length of the legs. Swimming on the surface, they can thus dip deeper to seize their prey. The same remark may be extended to the Saurians, in which, though the majority have a short neck, one fossil animal,[1] which appears to be the analogue of the swan, has a very long one. Other birds, as well as those of the present Order, are distinguished by the length of the neck ; as the peacock, the turkey, and several other Gallinaceans, and the Ostrich and its congeners are still more remarkable in that respect. This structure is probably as useful to them as to the web-footed birds, in enabling them to secure articles of food that would otherwise be out of their reach.

The birds at the foot of this Order, and indeed of the whole Class, are the *short-winged swimmers*, particularly the *auk*[2] and the *penguin ;*[3] the one having its station in the northern, and the other in the southern seas, reaching to the antarctic circle. The *northern* one, the *auk*, seems to rank above the penguin, for its wings have those feathers which, from their office being to propel birds when they fly, are denominated *rowing* feathers,[4] and they can flutter and flap their wings, while the *penguins* have

---

[1] *Plesiosaurus dolichodeirus.*  [2] *Alca.*
[3] *Aptenodytes.*  [4] *Remiges.*

none of these feathers, and cannot use their, so called, *wings* as such. The legs of the auk, also, are not placed quite so near the tail as in the southern bird, in which they are close to it, though both stand nearly in a vertical position. But though of no apparent use as *wings*, their short anterior appendages that go by that name, are not given them by their Creator merely for show, for when under water they use them as *fins;* and when it is recollected that Captain Beechy found them between three and four hundred miles from any land,[1] they seem to have occasion for additional rowing organs. One traveller, D. Pagès, says that they also sometimes use their wings as fore-legs, walking on all fours.[2] Some of them burrow like rabbits, but how they effect this has not been ascertained. In general they are reckoned as the most stupid and foolish animals in the whole Class: in fact most of the web-footed birds exhibit less of the life and spirit and gaiety that distinguish so conspicuously those whose principal theatre of motion is the air: belonging as they do to two elements, they may be regarded, in some sense, as half fowl and half fish; and when we call a man, not remarkable for sense, a *goose*, we admit some such degradation in aquatic birds.

[1] Voyage, i. 16.    [2] *N. D. D'H. N.* xiii. 306.

But all sea-birds are not of this character; amongst these the *frigate-bird*[1] and the *albatross*[2] are most conspicuous, emulating the eagle and the vulture amongst the terrestrial birds of prey. Of all the oceanic birds, the frigate-bird comes nearest to the *eagle*. Its keen sight, its crooked beak, its short, robust, and plumy legs, its sharp claws, the vast extent of its wings, and its rapid flight, all show that it is the oceanic representative of the king of birds. If the peaceful flying-fish seeks a refuge from the dorados[3] and bonitos,[4] its aquatic enemies, by elevating itself from the water into the air, the frigate-bird darts upon it like a thunder-bolt and devours it. If the booby[5] has caught a fish, like the bald eagle[6] the frigate-bird often compels it to let go its prey, and seizes it before it reaches the water. Its extent of flight is wonderful, and exceeds that of any other marine bird; for it possesses between the tropics a domain of more than four hundred leagues, over which it directs its course by day and by night; for, as the plumage of the under side of its body is not impervious to the water, it cannot continue long upon it, but prefers to brave the wind and

[1] *Tachypetes Aquila.*          [2] *Diomedea exulans.*
[3] *Coryphæna hippurus.*          [4] *Scomber Pelamis.*
[5] *Sula Bassana.*
[6] Richardson, *Fn. Boreal. Americ.* ii. 15.    Audubon. *Biogr.* 162.

the tempest, and to elevate itself above the storm, and for repose retires to lofty rocks and woody islets.

The *albatross* is the analogue of the *vulture*, and the largest of the sea-birds, and his wings expand sometimes to the extent of twenty feet; like his prototype, he is occasionally so gorged with food as to lose the power of flying, and when pursued, his only resource is to disgorge his overloaded stomach. Mr. Bennet has given a very interesting account of the mode of flight of this bird, to which I must refer the reader.[1]

I observed, in the last chapter, that one of the short-winged family of this Order, the *merganser*, appears to be connected with the *Saurians* by its serrated beak; but the *penguins*, which are at the foot of the same Family and of the Order, seem connected with the *Chelonians*, their rudimental wings and their legs approaching the paddles and webbed feet of the turtles and some of the tortoises. Their plumage, when not analyzed, resembles very much the fur of a seal, or some quadruped.

*Order* 2.—I have already noticed several circumstances relative to the birds of this Order;[2] I shall not, therefore, in this place, enlarge much upon them. Their general function is not only to devour the smaller fishes, aquatic

[1] *Wanderings*, &c. i. 45—47.
[2] See above, pp. 177, 194.

Molluscans, and other animals, as well as their spawn, that inhabit the waters of the globe, whether salt or fresh, but also those that are found in their vicinity, as worms, small reptiles, and insects in their different states; and their form is particularly adapted to their function: very long legs and toes; naked knees; a long sharp beak; where they have to dip under water for their food a long neck; and as, on account of their great length, they could not conveniently double their legs in flight, their tail is usually extremely short, so as to permit the legs to be stretched out, and act in some degree as steering organs. The body of these birds, generally speaking, in shape, seems to approach that of the *Scratchers*, but is rather longer, and not so plump. The form of some of them is very elegant and graceful; the plumage of others, especially of some of the scolopaceous tribe, is beautifully mottled, but, generally speaking, their colours are not brilliant.

There is one bird[1] of this Order that is particularly interesting, not only on account of some singularities in its structure, but likewise for its amiable manners: this bird is described and figured by Piso[2] under the name of *Anhyma*, but it is more commonly known by that of *Kamichi*,

---

[1] *Palamedea cornuta.*

[2] *Hist. Nat. et Med. Ind. Occid.* 91.

It is said to be larger than the peacock or even the swan. Its wings are armed with two strong spurs, which point outwards when the wing is folded; but its most remarkable feature is the long, slender, cylindrical, and nearly straight horn which arms its forehead. One would suppose a bird so fitted for combats was the terror of the feathered race, delighting in battle and bloodshed, but this is not the case, for it is one of the most gentle and susceptible of birds. It feeds upon grass, and attacks no birds that approach it: at the time of pairing, however, the males contend fiercely and sometimes fatally for the females; but the victory gained, they become patterns of conjugal fidelity, never parting, and like the turtle, if one outlives the other, the survivor usually is the victim of its grief.[1]

Another South American bird of this Order,[2] if we may credit the accounts that are given of it, is gifted by its Creator with an instinct still more wonderful; it seems to have a natural inclination for the society of man, and seems to occupy the same place amongst birds that the *dog* does amongst quadrupeds. When taken and fed in a house, it becomes attached to the inmates. Like the dog it knows the voice of its master, and will follow or precede him when he

---

[1] Sonnini, in *N. D. D'H. N.* xvii. 21.

[2] *Psophia crepitans.*

goes out, quits him with reluctance, and appears delighted when it sees him again. Sensible of his caresses, it returns them with every mark of affection and gratitude: it seems even jealous of his attentions, for it will peck at the legs of those who come too near to him. It knows and acknowledges also the friends of the family. It sometimes takes a dislike to individuals, and whenever they appear, attacks them, and endeavours to drive them away. Its courage is equal to that of the dog, for it will attack animals bigger and better armed than itself. Sonnini, who relates the preceding anecdotes from his own observation, was also told that in some parts of America, these birds were entrusted with the care of the young poultry, and even of the flocks of sheep, which they conducted to and from their pastures.[1]

The *common Stork*[2] seems equally attached to man, and in return has generally met with protection from him, and in many nations has been accounted a sacred bird that it is a sin to kill or molest; and they are entitled to these immunities not only on account of their philanthropic instincts, but likewise because they destroy lizards, frogs, serpents, and other noxious reptiles, which are a considerable annoyance in low and marshy districts. The *black Stork*[3] is

---

[1] *N. D. D'H. N.* i. 190.   [2] *Ciconia alba.*   [3] *C. nigra.*

of a less social turn, and avoids the neighbourhood of man, and frequents solitary marshes
and thick woods, where it nidificates on old
trees.

*Order* 3.—We seem to enter this order—which
from the swiftness of the few animals that compose it, is called the Order of *Coursers*[1]—by one
of the most singular birds that is at present
known; I mean the *Apteryx australis* of Dr.
Shaw. As far as can be judged from the only
known specimen, which was brought from New
Zealand in 1812, one would think this bird
*osculant* between the *Waders* and the present
Order. Its legs, indeed, seem those of a gallinaceous bird, with a tendency, as Mr. Yarrel
remarks, to the spurs of that tribe,[2] but its beak
is related to that of the *Ibis*, and the lateral skin
of the toes is notched as in the *Phaleropes*.
The wings are shorter than in any other known
bird, quite concealed by the feathers, and terminate in a *claw;* a circumstance which seems to
indicate an approximation to some quadruped
form. These wings, though useless for flight,
were doubtless given by its Creator to this
animal to answer some purpose in its economy,
either as a weapon or a prehensive organ. With
the birds of the Order in which it is placed it
agrees in its general form and plumage, but in

---

[1] *Cursores.*       [2] See *Zool. Trans.* i. i. t. x. 74.

stature it falls below them, being of the size of a small turkey. It is called by the natives *Kivi*.

There is another insular bird, the *Dodo*, noticed in a former chapter,[1] which though classed with *this*, to judge from its figure seems to connect the Ostrich with the *next* Order, the *Scratchers*;[2] but if we suppose the Order to form a *circle*, these birds will meet, one still being conterminous to the Order above it, and the other to that below it. These two birds have *four* toes. Mr. W. S. Mac Leay,[3] as well as several other zoologists, is of opinion that the Ostrich Family, meaning the typical members of it, both in their *internal* as well as their *external* structure, approach the nearest to Mammalians. Of the Ostrich itself it is stated, amongst other characters, that its upper eyelid is moveable and ciliated, and that its eyes are more like the eyes of a man than those of a bird, and they are so set as both of them to see the same object at the same time; that it is the only bird that discharges urine,[4] with many circumstances which I have no room to enumerate. Mr. Owen, however, whose accuracy as a comparative anatomist can be fully relied on, has observed to me, that the urinary bladder, sternum, and some other

[1] Vol. I. p. 55.  [2] Vigors in *Linn. Trans.* xiv. 485.
[3] *Hor. Ent.* 266.  *Linn. Trans.* xvi. 43.
[4] *N. D. D'H. N.* iii. 85, 86.

parts of these birds, are closer approximations to the *Chelonians* than the *Mammalians*.

The animal of the latter Class, whose external form approaches nearest to the Ostrich is the *Camel*, a resemblance which has been so striking, that from a very early period they have been designated by a name which connects them with this quadruped:[1] in many particular points, besides general form, they also resemble it. The substance and form of their two-toed feet, a callosity on their breast and at the os pubis, their flattened sternum, and their mode of reclining. It is singular that these birds associate with beasts, particularly the quagga and zebra.[2]

The new world, which has a representative of the camel in the *lama*, and of the hippopotamus in the *tapir*, has also a peculiar *ostrich* of its own, which is called the *nandu*;[3] so that in Africa, Asia,[4] Australia,[5] and America, there is a distinct genus of the present Order, each, as at present known, consisting of a single species.

With respect to their *functions*, not much has been observed: they are said to live a good deal upon grain, fruit, and other vegetable substances, and the nandu is fond of insects; probably others of them may also assist in restraining the in-

---

[1] *Struthio-camelus.*

[2] Burchell's *Travels in S. Africa*, ii. 315.

[3] *Rhea americana.*     [4] *Casuarius geleatus.*

[5] *Dromaius ater.*

cessant multiplication of these little creatures. The ostrich may be said almost to graze, though it is very eager after grain ; but its history is too well known to require any further enlargement upon it.

*Order* 4.—The birds of this Order are called *Scratchers,* from an action common to many of them, and more particularly observable in our common poultry, that of *scratching* the ground to turn up food, especially when followed by their chicks. Of all the gifts of Providence, there is none that more promotes our comfort and pleasure than the majority of the animals that compose this Order, for it includes almost all our barn-door fowls, and the great majority of the game pursued so eagerly by the sportsman : birds not only valuable for the variety and delicacy of the food, both flesh and eggs, with which they supply our tables, but delighting us by the beauty, the elegance, and stateliness of their forms ; the diversity of their plumage, especially the elongated or expansile tail feathers of the males; and the rich variety and splendor of their colours. The gorgeous peacock and the graceful pheasant have scarcely a parallel in the other Orders, except perhaps, as to splendor, in those brilliant little gems, the humming birds.

I have mentioned, on a former occasion,[1] the numerous varieties of the common fowl, which

[1] VOL. I. p. 66.

have probably been produced by climate and cultivation. With regard to size, Sumatra appears to produce both the smallest and the largest kind of poultry, the common feather-legged *Bantam*, and the *Iago* fowl,[1] the cock of which, Marsden says, he has " seen peck off a common dining table; when fatigued, they sit down on the first joint of the leg, and are then taller than the common breed."[2] Colonel Sykes imported them into England in 1831; the hen laid freely, and reared two broods of chickens.

Wild poultry are found both in the old world and the new: the jungle-fowl,[3] from which our breeds are supposed by Sonnerat to have originated, are common in India; and the Spaniards are said to have found another kind in Peru and Mexico, in which last country they were domesticated, and called *chiacchialacca;* Parmentier states that he heard the crow of the cock of this breed in the wildest forests of Guiana, and that he had seen one of them.[4]

The Birds of this Order are granivorous, insectivorous, or both, and the Hocco is stated to subsist on buds and fruits. Some are gregarious,

---

[1] *Gallus giganteus.*   [2] *Sumatra,* 2 Ed. 98.
[3] *Gallus Sonneratii.*
[4] *N. D. D'H. N.* vii. 472. Modern ornithologists appear to account all these breeds as well as those mentioned in a former chapter (VOL. I. p. 66) as distinct species. Linné, besides his *Phasianus Gallus a,* or the common breed, has Var. β, *P. G.*

as the pigeons; while others, as the partridge, form coveys only for a time; in spring those that survive the sporting season pair off, and are soon at the head of a numerous family.

*Order* 5.—Baron Cuvier has separated the *Climbers* from Mr. Vigor's Order of *Perchers*, not only on account of their having two toes behind, as well as before, but also on account of differences in their larynx, sternum, and cœcal appendages. Amongst the Climbers, though there are some armed with beaks of very extraordinary forms and magnitude, as the toucan, there are none so interesting and altogether so remarkable as the Psittacean Family, or the Parrots, Parroquets, Macaws, Cockatoos, &c. They seem complete analogues of the Monkeys and other Quadrumanes, which they exceed, in their faculty of learning to articulate many words, for which their lower larynx is particulary constructed, and thus mimic the *utterance* of man, as the former animals do his *actions;* a circumstance which seems to have induced some ornithologists to place them at the head of their Class,[1] in contrast with the latter animals.

---

*cristatus,* or the *Polish* breed; γ. *P. G. ecaudatus,* or the *Rumplet;* δ. *P. G. Morio,* or the *black-skinned* breed; ε. *P. G. lanatus,* or the *silk* breed; η. *P. G. crispus,* or the *Friesland* breed; and ζ. *P. G. pusillus,* or the *Bantam* breed. There are several more in Gmelin.

[1] Illiger, &c.

There is a genus, belonging to this Order, found in the southern parts of Africa, the species of which are called *bee-cuckows*,[1] and are remarkable for indicating both to the honey-ratel[2] and the Hottentot the subterranean nests of certain bees, which they do by a particular cry, morning and evening, and by a gradual and slow flight towards the quarter where the swarm of bees have taken up their abode; the beast and the man both attend to the notice, seek the spot, and dig up the nest; and to the share of the bird generally falls, not the part stored with the *honey*, but that in which the *grubs* are contained:[3] so that the bird, though it invites others to partake with it, has its own subsistence, which it could not otherwise readily come at, principally in view. Both this animal and its companion, the ratel, are fitted by Providence for their function, and protected from the danger to which they are exposed from the strings of the irritated bees by a very hard skin. The bees, however, sometimes revenge themselves on the treacherous bird by attacking it about the head and eyes, and so destroying it.[4] It is singular, and affords a most convincing proof of design, that two animals that are so necessary to each other, the

---

[1] *Indicator major, minor Vieill*, &c.

[2] *Vivera mellivora.*

[3] Sparrmann, *Voyage*, ii. 181, 187.

[4] Cuv. *Règn. An.* i. 455. Sparrmann, *Voy.* ii. 182.

one to indicate and the other to excavate their common prey, should each be defended by the same kind of armour, and each seek a different portion of the spoil, suited to its habits.

Amongst the birds most remarkable for their instincts, in the present Order, is the *wryneck*.[1] It is a feathered *ant-eater*, and is organized by its Creator to entrap its prey by the very same means as the quadruped ones. Like them, it can protrude its tongue to a very great length, which is not owing to the structure of this organ itself, but to a peculiar ligamentous sheath in which it usually is contained. Its salivary glands are above an inch long, and shaped somewhat like a tea-spoon. The saliva they secrete is so very viscid as to be capable of being drawn into threads finer than a hair, and several feet in length; so that when the tongue is besmeared with it, no insect that touches it can escape. Like its analogues, it darts its tongue into an ant-hill, or lays it on an ant-track, and draws it back into its mouth laden with prey.[2] It is singular that the functions, in warm climates, given in charge by Providence to. *quadrupeds*, in tem-

[1] *Yunx torquilla.*

[2] I owe these observations on the wryneck principally to a medical friend, George Helsham, Esq. of Woodbridge, in Suffolk, a practical ornithologist, not only systematically and anatomically, but knowing birds also in their haunts, and conversant with their habits and instincts.

perate ones, in this instance, devolves upon
*birds*. The rapid increase of ants, in tropical
countries, probably rendered it necessary that
their devourers should be more numerous, and
act with a greater momentum.

The general functions of this Order, as they
are in most of those of the present Class, are
various. The food of some are roots, fruits,
and other vegetable substances;[1] of others the
grubs of insects;[2] of others, again, principally
insects in general under every form;[3] and
lastly, some to fruits or insects will add the
eggs and the nestlings of other birds.[4]

*Order* 6. — The birds of this Order, the
*Perchers*, are distinguished from the last, not
only by the characters lately noticed, but likewise by a considerable difference in their habits
and manners. Amongst them we find all those
that delight us by their varied song; they are
truly birds of the *air*, for they seem to have the
full command of that element; many of them
moving gaily in every direction that their will
suggests, rising and falling, flying backwards
and forwards, or performing endless evolutions,
*pro re nata*, in their flight. These *Perchers* also
are the best nest-builders, not usually selecting,
like the Climbers, the interior of a hollow tree

---

[1] The *Psittaceans.*      [2] The *Pies.*
[3] The *Cuckows.*         [4] The *Toucan.*

or similar situations, but most commonly interweaving their nests between the twigs and
branches of trees and shrubs, or suspending
them from them, or even attaching them to
humbler vegetables; some having even exercised arts from the creation, which man has
found of the greatest benefit to him, since he
discovered them. These birds, indeed, may be
called the inventors of the several arts of the
weaver, the sempstress, and the tailor, whence
some of them have been denominated *weaver* and
*tailor-birds.*

The nest of the little Indian *weaver-bird*,[1]
though it has neither warp nor woof, being
formed by various convolutions of the slender
leaves of some grass, so intertwined and entangled as to produce a web sufficiently sub-
stantial for the protection of the inhabitants of
the nest, is, nevertheless, a very wonderful
structure, but as it is well known[2] I shall not
further enlarge upon it, but proceed to the
*tailor-birds*, whose nests are still more remarkable.

India produces several species that are instructed by their Creator to *sew* together leaves
for the protection of their eggs and nestlings
from the voracity of serpents and apes; they

---

[1] *Ploceus Textor.*

[2] There are several of these nests in the museum of the Zoological Society.

generally select those at the end of a branch
or twig, and sew them with cotton, thread, and
fibres. Colonel Sykes has seen some in which
the thread was literally knotted at the end.[1]
The Indian birds of this description form two
genera, separated from *Sylvia* by Dr. Horsfield.[2]
The inside of these nests is lined usually with
down and cotton.

But these birds are not confined to India or
tropical countries ; Italy can boast a species
which exercises the same art : and I am indebted
to the kindness of one of our most eminent ornithologists [3] for being enabled to give a figure of
this pretty and interesting bird, from a specimen
in his possession ;[4] and to the Zoological Society
for their permission to have a drawing made from
a nest in their museum.[5] This little creature
was originally described and figured by M. Tem-
minck in 1820, but its singular instincts, as to its
mode of nidification, were afterwards given in
detail by Professor P. Savi. It is called by the
Pisans *Becca moschino*, and is a species of the
genus *Sylvia*.[6]

In summer and autumn it frequents marshes,
but in the spring it seeks the meadows and corn-
fields ; in which, at that season, the marshes
being bare of the sedges which cover them in

---

[1] *Catalogue of birds*, &c. 16.    [2] *Prinia* and *Orthotomus*.
[3] Mr. Gould.    [4] PLATE XV. FIG. 1.
[5] *Ibid.* FIG. 2.    [6] *S. cisticola*.

the summer, it is compelled to construct its
nest in tussocks of grass on the brink of ditches:
but the leaves of these, being weak, easily split,
so that it is difficult for our little sempstress
to unite them, and so to form the skeleton of
her fabric. From this and other circumstances
the *vernal* nests of these birds differ so widely
from those made in the autumn, that it seems
next to impossible that both should be the work
of the same artisan.

The latter are constructed in a thick bunch
of sedge or reed, they are shaped like a pear,
being dilated below and narrowed above,[1] so as
to leave an aperture sufficient for the ingress
and egress of the bird. The greatest horizontal
diameter of the nest is about two inches and
a half, and the vertical is five inches or a little
more.

The most wonderful thing in the construction
of these nests is the method to which the little
bird has recourse to keep the living leaves
united, of which it is composed. The sole
interweaving, more or less delicate, of homogeneous or heterogeneous substances forms the
principal adopted by other birds to bind together
the parietes of their nests; but this *Sylvia* is no
weaver, for the leaves of the sedges or reeds
are united by real *stitches*. In the edge of each

---

[1] PLATE XV. FIG. 2.

leaf she makes, probably with her beak, minute apertures, through which she contrives to pass, perhaps by means of the same organ, one or more cords formed of spiders' web, particularly of that of their egg-pouches. These threads are not very long, and are sufficient only to pass two or three times from one leaf to another; they are of unequal thickness, and have knots scattered here and there, which in some places divide into two or three branches.

This is the manner in which the exterior of the nest is formed; the interior consists solely of down, chiefly from plants, a little spiders' web being intermixed, which helps to keep the other substances together. In the upper part and sides of the nest, the two walls, that is the external and internal, are in immediate contact; but in the lower part a greater space intervenes, filled with the slender foliage of grasses, the florets of Syngenesious plants, and other materials which render soft and warm the bed in which the eggs are to repose.

This little bird feeds upon insects. Its flight is not rectilinear, but consists of many curves, with their concavity upwards. These curves equal in number the strokes of the wing, and at every stroke its whistle is heard, the intervals of which correspond with the rapidity of its flight.

Perhaps of all the instincts of Birds, those

connected with their nidification are most remarkable; and of all these, none are so wonderful as those of the tribe to which the little bird
whose proceedings in constructing its nest I
have just described, belongs. In the Indian
tailor-birds, the object of their sutorial art is
stated above; and doubtless, in the case of the
Italian, the attack of some enemy is prevented
by her mode of fabricating her nest. Situated
so near the ground, her eggs, but for this
defence, might otherwise become the prey,
perhaps, of some small quadruped or reptile.
He who created the birds of the air taught every
one its own lesson, and how to place and construct its nest as to be most secure from inimical
intrusion. I may observe here, that Professor
Nitzch's three Orders, or rather Sub-classes,
mentioned above, receive some confirmation
from the places selected by the individuals com-
posing them, to form their nests and deposit
their eggs in. The *aquatic* birds generally select
places in the vicinity of *water;* the *terrestrial*
make them on the *ground;* and the great body
of the *aërial* construct their nests in *trees, shrubs,*
and *plants.*

The birds of this Order as to their *food* leave
no vegetable or animal substance untouched, and
the humming-birds, with their butterfly-tongue,
imbibe the nectar of flowers. Of a vast number,
insects form the principal part of their food, and

they are the chief check to their too great multiplication; and sometimes, as in the case of
the locust-eating thrush,[1] they devote themselves
to a particular tribe of insects, but most of the
insectivorous birds will also eat grain.

*Order* 7. — The last Order of Birds, the
*Raveners*, includes those that are most perfect
in their form, and all are remarkable for their
predatory habits. Their power of wing, and
talon, and beak, distinguish them from all other
birds of the *air;* and though some of the terrestrial birds vie with them in magnitude, and
some of the aquatic ones, as we have seen,[2]
exceed them in extent of wing and untired
flight, yet none can come near them in the
union of all those qualities which constitute their
claim to the first rank amongst the birds; and
the eagle has, as it were, been *consecrated* king
over them all, by being placed in the Holy of
Holies of the Jewish temple as one of the
symbols of those powers that rule under God in
nature.[3]

This Order is usually divided into two sections, which might be denominated Sub-orders,
the *nocturnal* birds of prey and the *diurnal*.
The first of the birds of these sections are distinguished by their large eyes, the enormous pupil

[1] *Turdus gryllivorus.*      [2] See Introd. lxxii.
[3] Ezek. i. 10; x. 1.

of which receives so many rays of light, that they are dazzled by the glare of day; but by it are enabled to see in the night—they fly in the evening and by moonlight. Thus they are fitted best to fulfil their function, and to be very beneficial to man, in keeping within due limits animals that are often extremely detrimental to his property, and commit their ravages more or less in the night; on this account owls are often seen in barns where mice and rats abound, and are most valuable auxiliaries to the cats. The white owl[1] is said to destroy more of the murine race than even these last animals. Had not the provident care of the Father of the universe created these mouse-and-rat-destroying animals, the tiller of the soil would often labour in vain.

The *diurnal* Section of the Raveners contains all the birds of might and power. I have before mentioned the secretary bird,[2] created to diminish the number of serpents; so similar to some of the *waders*, as to have been classed with them by several ornithologists; but Cuvier says, its whole anatomical structure, as well as its beak and other external characters, vindicate its claim to be placed in the present Order.[3]

Another species belonging to it descends to still lower food, and like the bee-eater,[4] devours

---

[1] *Strix flammea.*  
[2] See above, p. 178.  
[3] *Règne An.* i. 339.  
[4] *Merops apiaster.*

bees and wasps and other insects, I allude to the
*bee-falcon;*[1] but in general the aquiline race
attack vertebrated animals, reptiles, fishes, and
birds of every wing, and many quadrupeds, and
the giant vultures satiate their ravenous appetites
upon any carcases that their piercing sight, from
the great heights to which they ascend, can
discover. Humboldt says, that the Condor[2]
soars to the height of Chimborazo, an elevation
almost six times greater than that at which the
clouds that overshadow our plains are suspended.[3]

In the book of Deuteronomy we have a very
animated and beautiful allusion to the eagle, and
her method of exciting her eaglets to attempt
their first flight, in that sublime and highly
mystic composition called Moses' Song ; in which
Jehovah's care of his people, and methods of
instructing them how to aim at and attain
heavenly objects, is compared to her proceedings
upon that occasion. *As an eagle stirreth up her*
*nest, fluttereth over her young, spreadeth abroad*
*her wings, taketh them, beareth them on her wings:*
*so Jehovah alone did lead him.* The Hebrew
lawgiver is speaking of their leaving their eyrie.
Sir H. Davy had an opportunity of witnessing
the proceedings of an eagle after they had left
it. He thus describes them.

---

[1] *Pternis apivorus.*       [2] *Sarcorhamphus Gryphus.*

[3] *Zool.* i. 29.  See above, p. 155.

" I once saw a very interesting sight above one of the crags of Ben Nevis, as I was going on the 20th of August in the pursuit of black game. Two parent eagles were teaching their offspring, two young birds, the manœuvres of flight. They began by rising from the top of a mountain in the eye of the sun; it was about mid-day, and bright for this climate. They at first made small circles, and the young birds imitated them; they paused on their wings, waiting till they had made their first flight, and then took a second and larger gyration, always rising towards the sun, and enlarging their circle of flight so as to make a gradually extending spiral. The young ones still slowly followed, apparently flying better as they mounted; and they continued this sublime kind of exercise, always rising, till they became mere points in the air, and the young ones were lost and afterwards their parents to our aching sight."[1]

What an instructive lesson to Christian parents does this history read! how powerfully does it excite them to teach their children betimes to look toward heaven and the Sun of righteousness, and to elevate their thoughts thither more and more on the wings of faith and love; themselves all the while going before them, and encouraging them by their own example.

[1] Salmonia, 99.

# Chapter XXIV.

## *Functions and Instincts.   Mammalians.*

WE are now arrived at the last and highest Class of the Animal Kingdom, to which man himself belongs, and of which he forms the summit: but though he may be said to belong to it in some respects, in others he stands aloof from it, as an insulated animal, and one exalted far above it, being created rather to govern its members, than to be the associate of the highest of them.

This Class includes many animals which are of the greatest utility to man, and without which he could scarcely exist, at least not in comfort; and others again that attack him and his property; and though the fear of him, in some degree, still remains upon them, also often excite that passion in his breast.   But he of all animals is the only one, that by the exercise of his reasoning powers and faculties, can arm himself with factitious weapons enabling him to cope with the superior strength, the fierceness, claws, and teeth of the tiger or the lion, and to lay them dead at his feet when in the very act of springing upon him.

The animals of this Class, that are *terrestrial*, are all *quadrupeds*,[1] and are mostly covered with fur or hair, longer or shorter, though in some, these hairs become quills, as in the porcupine, or spines, as in the hedgehog; others, like the serpents and lizards, are protected by scales, as the *Manis;* and some are incased in a hard coat of armour, often consisting of pieces so united as to form a kind of mosaic, as the armadillo, the *Chlamyphorus*,[2] and probably the *Megatherium*.

In the *aquatic* Mammalians the legs are, more or less, converted into *fins*, or means of natation.[3] The whole body constituting the Class, though sometimes varying in the manner, are all distinguished by *giving suck* to their young, on which account they were denominated by the Swedish naturalist, *Mammalians.*[4]

The situation and number of the, usually protuberant, organs that yield the milk, vary in different tribes and genera. The Creator has distributed them according to the circumstances of each kind. Physiologists divide them into *pectoral*, or those on the *chest; abdominal*, or those on the *abdomen;* and *inguinal*, or those on the *groin.* In the human race, the *Quadrumanes*, and the *bats*, and some others, these organs are

---

[1] Τετραποδα της γης.       [2] PLATE XVII.

[3] See above, p. 126. 143.

[4] Cuvier calls them *Mammifera*, but there seems no reason for altering the original term.

placed between the arms.  For an erect animal like man, it is evident that this situation for the paps was the only convenient one for suckling an infant, either when sitting or standing ; the monkey tribes also, which are always moving about upon trees, and among the branches, could not have exercised this maternal function, had their lactescent organs been placed lower ; and the bats, which carry and suckle their young during flight, required that their nipples should be similarly placed, to enable them to keep fast hold. All the species of the above tribes have only a *pair* of the organs in question, with the exception of the lory, or sloth-ape,[1] so called from the excessive slowness of its movements, which has *four*, two of which Cuvier places in his abdominal column, under the name of *epigastric*.

The animals which produce more than two at a birth, as might be expected, have a proportionable number of nipples differently distributed. Thus the *cat* has four pectoral, and four abdominal.  The ten nipples of the *swine* are all abdominal, and those of the other Pachyderms, with the exception of the elephant, which has only two pectoral nipples, are similarly situated. The jerboa[2] has both pectoral and inguinal ones, while the lemming[3] has all three kinds ; the *Ruminants, Solipeds, Amphibians, Carnivorous*

---

[1] *Stenops.*        [2] *Dipus Sagitta.*        [3] *Lemmus.*

*Cetaceans,* have only inguinal dugs, with from two to five nipples. This situation is evidently best suited for suckling their limited number of young ones. Amongst the *Marsupians,* whose young, immediately upon their birth, pass into a second matrix as it were, almost the entire skin of the abdomen forms a pocket, inclosing the lactescent organs; those of the opossum are arranged, in Cuvier's table, in the inguinal column ; but in the *Kanguroo,* which has four, they appear rather to be abdominal. These variations in the position and number of the organs furnishing the sole food of the animals of the present Class in their state of infancy, were evidently planned and formed by the hand of a Being supreme in Wisdom, Power, and Goodness, who adapted every organ to the circumstances in which it was his will to place the diversified animals that compose it, and to their general structure. To those which produce not more than two at a birth, only two organs for suction were usually given, placed, according to the wants of the animal, either between the anterior or posterior extremities, in which latter case the posture was never erect; but where he decreed an animal should produce a more numerous progeny, he planted them in greater numbers, and so distributed them that all belonging to the same litter could suck at the same time. In the case of the Kanguroo the members of *two* litters are sometimes sucking at the same time, which accounts for their having *four* nipples, a fact which shows how accurately every thing has been foreseen, weighed, and numbered, by a Provident Intellect.

In the whole animal kingdom, except amongst the Mammalians, there is no instance of the young being supported by their parents with nutriment derived from themselves, nothing, therefore, affords a clearer character for a definition of the Class than this most interesting one : the Birds, indeed—with the exception of pigeons which feed their nestlings from their crop—as well as the bees, and several other Hymenopterous insects, provide their progeny with food which they collect for them themselves ; but the great majority of invertebrated animals, confine their care for them, to placing their eggs, in a situation in which, when hatched, they would meet with their appropriate food, and this appears to be all that is generally done by the two first classes of Vertebrates, the Fishes, and the Reptiles.

## MAMMALIA. *(Beasts.)*

*Animal* vertebrated, ovoviviparous, or viviparous.

*Extremities* ambulatory, or natatory ; in a few organized for flight.

*Integument* pilose; sometimes spinose, or armed with hard scales or plates; and sometimes naked. *Young* not hatched by incubation, but when first extruded from the matrix, receiving their nutriment by suction, till they can support themselves.

*Circulation* double. *Blood* red, warm.

*Respiration* simple. *Lungs* thoracic.

Cuvier seems to have laboured under some difficulty with regard to the *Classification* of Mammalians, and to have regarded the Marsupians and Monotrèmes as forming a *distinct* Class, divisible, for the most part, into Orders analogous to those into which the Class of common Quadrupeds is divisible.[1] Subsequent observations have proved the general correctness of this idea. Mr. Owen observes to me, in a letter, " Dissections of most of the genera of *Marsupians* have tended to confirm in my mind the propriety of establishing them as a distinct and parallel group, beginning with the *Monotrèmes*, which I believe to lead from *Reptiles*, not birds. A general simplicity in the structure of the brain; a less perfect condition of the vocal organs; some peculiar dispositions of the great veins and arteries, as the presence of two superior *venæ cavæ*, and the absence of an *inferior mesenteric artery*, are among the circumstances in which they, the

---

[1] *Règn. An.* i. 174.

Marsupians and Monotremes differ from the true viviparous Mammalians, and agree with the oviparous Vertebrates. Recent opportunities of examining the impregnated uterus of the *Kanguroo* and *Ornithorhynchus* have almost determined that they are both ovoviviparous."

Under these impressions, confirmed and illustrated by the observations of so able a comparative anatomist, I shall consider the Class of Mammalians as divisible into two *Subclasses*, viz. *Ovoviviparous* Mammalians, and *Viviparous* Mammalians.

It may be here observed, with regard to the state of forwardness in which the different tribes of Mammalians leave the matrix, a considerable variation takes place, some requiring a longer time than others, before they can be considered as at all independent of maternal care and protection. The young of the Ruminants, Pachyderms, and Solipeds, come into the world with the organs of the senses, and of locomotion, in a state to be used immediately; they can *see* with their *eyes,* and *hear* with their *ears,* and *walk* with their *legs,* as soon as they are born; whereas the Predaceans and several others, when first born are *blind,* and unable to *walk,* and do not attain to the full use of their eyes and legs till a considerable time after birth. In man, though the infant is born *seeing,* yet a much longer period, and the instruction of the mother or nurse, are required before it can *walk*.

In the first case here noticed, that of the Ruminants and Pachyderms, the young animal requires less care from the mother. She has little to do besides suckling, and watching it in order to protect it if danger threatens. But, in the second case, she must prepare a kind of nest, not exposed to the light, and removed from observation, in which she can attend to her young unmolested, till they can see and move about upon their legs. Every one knows how attentive feline animals are to these circumstances, and the Rodents often excavate burrows in which they bring forth and suckle their young. The Marsupian Mammalians probably are exposed to external circumstances, which render it necessary that they should have a kind of nidus formed of the skin of their own body, to receive their young when they leave the matrix, at which period they seem to be in a more helpless state than any of the animals last alluded to.[1]

From this statement we see that the graminivorous and omnivorous animals, whose food is always at hand, come into the world the best prepared for action ; while the carnivorous ones, and those that must, if I may so speak, procure their daily bread by the sweat of their brows,

---

[1] Owen in *Philos.* Tr. 1834. 344.

require to be in some degree *educated* for their function,[1] before they can duly exercise it. In the instance of the *Ornithorhynchus*, a burrow[2] seems to supply the place of the marsupial pouch, which indicates some approach to many of the Rodents.

*Sub-class* 1. *Ovoviviparous* Mammalians.

*Chorion*, or external membrane of the egg not rendered vascular by the extension of the fœtal vessels into it. *Embryo* not adhering to the uterus.

Only one *passage* out of the body.

*Marsupial bones* in all.

This Sub-class is divided into two Orders, *Monotremes*, and *Marsupians*.

*Order* 1. — *Monotremes* (*Ornithorhynchus; Echidna.*)

No marsupial *pouch*. *Coracoid bones* extended to the sternum. Young suckled from a *mammary orifice:* brought up in *burrows*. *Animal* predaceous.

*Order* 2. — *Marsupians* (*Wombat; Koala; Kanguroo; Phalangist; Flying* and *Common Opossum*, &c.)

A marsupial *pouch* receiving the young after birth, in which they are suckled, by means of *nipples*. *Animal* herbivorous, predaceous, or carnivorous.

---

[1] See above, p. 261.   [2] Owen, *ubi supr.* 564.

*Sub-class* 2.—*Viviparous* Mammalians.

*Chorion*, or external membrane of the *egg* rendered vascular by the extension of the fœtal vessels into it.

*Embryo* adhering to the uterus.

*Young* when brought forth not received into a pouch; suckled by a nipple.

This sub-class is divided into *eight* Orders thus arranged in an ascending scale,

1. *Cetaceans.*
2. *Pachyderms.*
3. *Ruminants.*
4. *Edentates.*
5. *Rodents.*
6. *Predaceans.*
7. *Cheiropterans.*
8. *Quadrumanes.*

Several of these *Orders* may be further divided into *Sub-orders*, as will appear when I come to treat of them. I have not adhered to Baron Cuvier's arrangement, in placing the *Ruminants* next to the *Cetaceans*, for it always appeared to me incongruous to place at the foot of the scale, animals on every account entitled to rank higher: and I am happy to find my opinion backed by Mr. Owen's judgment, which he informs me is grounded on anatomical considerations. The *Hippopotamus* appears to us both the proper successor of the Cetaceans.

*Order* 1.—*Cetaceans*. This Order may be divided into two *Sub-orders*, the *first* consisting of those that form the great body of the Order, which are *predaceous* in their habits; and the *second* of those that are *herbivorous*. (To the first belong the *Whales;* the *Cachalots;* the

*Narwhals;* the *Porpoises;* and the *Dolphins,*
&c.: and to the second, the *Manatee;* the
*Dugong;* and *Rytina.*)

This Order is principally distinguished from
the terrestrial Mammalians by having the *hind
legs* converted into a horizontal (so called) fin
moving up and down. They have little or no
neck, and their anterior extremities are covered
with a tendinous membrane, which enables the
animal to use them as fins.

The *Predaceous* Cetaceans are distinguished
from the *Herbivorous* by having their *mammary*
organs *inguinal,* and by their *fins* not being
prehensory.

In the *Herbivorous* Sub-order, the *mammary*
organs are *pectoral,* and they can use their *anterior* extremities, in some degree, as hands, to
carry their young, and in locomotion.[1] They
are also armed with *tusks,* a circumstance which
appears to connect them with the *Morse* or
*Walrus,*[2] which is said, by Cuvier, to be both
herbivorous and carnivorous, and to differ considerably from the rest of the *Amphibians.*

*Order 2.—Pachyderms.* The external characters which distinguish the *Solipeds* from the
*typical* Pachyderms are so striking, that they
seem almost entitled to be placed in a separate
Order. I shall, however, consider them as forming a Sub-order. (To this Order belong the

---

[1] See above, p. 136.          [2] *Trichecus rosmarus.*

*Hippopotamus* ; the *Tapir* ; the *Swine* tribe ; the *Rhinoceros* ; the *Elephant* ; the *Horse* ; and the *Ass* ; &c.) The principal characters of this Order, are *Feet* armed with *hoofs* incapable of prehension. In the *typical* Pachyderms the hoof is divided more or less, but in the *Solipeds* it is not.

*Order 3.—Ruminants.* The *Camel* tribe seems to form another Sub-order in the present Sub-class, distinguished by the remarkable circumstance, mentioned upon a former occasion, that its hoof, though superficially divided, has an entire sole,[1] and the males have no horns. (This Order includes the *Camel* ; *Dromedary* ; *Lama* ; *Giraffe* ; the *Ox*, and *Sheep* tribes ; the *Goats* ; the *Antelopes* ; the *Deers* ; and the *Elk*.) The principal character of the Order is that which its name indicates, that the animals belonging to it, *chew the cud*, that is, masticate a second time the food that they swallow, which, owing to the structure of their stomachs, they can return to the mouth after the first deglutition.

*Order 4.—Edentates.* (This Order contains the *Pangolin* ; the *Ant-eaters* ; the *Armadillos* ; and the *Sloths* ; &c.) Their distinctive character is to have no fore teeth.

*Order 5.—Rodents.* (*Guinea-pigs* ; *Hare* and *Rabbit* ; *Porcupine* ; *Beaver* ; *Mouse* ; *Rat* ; *Dor-*

---

[1] See above, p. 203.

*mouse; Jerboa; Marmot; Squirrels;* &c.) The principal character of this order are its *front* or *cutting teeth;* of these there are *two* in each jaw, separated from the grinders by an *interval,* so that they can neither seize any living prey, or lacerate its flesh; they cannot even cut the aliments which form their subsistence, but they can, as it were, file them, and by constant labour, nibbling and gnawing, reduce them to fragments proper for deglutition. They are connected with the *kanguroo,* the *wombat,* and other *Marsupians,* and the *beaver* exhibits one of the distinctive characters of the *Monotremes,* it has only *one* passage by which the *excrements* are ejected.

*Order 6.—Predaceans* or *Zoophagans.* Cuvier's subdivisions of this Order may be regarded, for the most part, as Sub-orders, but there is one tribe included in it by this great man, the *Cheiropterans,* which seems rather to form an Osculant Order, between it and the Quadrumanes. *(Walrus; Seals; Cat; Leopard; Panther; Tiger; Lion; Hyæna; Ichneumon, Civet-cat; Fox; Wolf; Dog; Otter; Martin; Weasel; Glutton; Bear; Mole; Hedgehog; Shrew;* &c.) The animals of this Order have *three* kinds of *teeth,* viz. *cutting-teeth, canine* teeth, and *grinders;* their *paws* are armed with *claws;* their muzzle is often set with *whiskers,* usually called smellers; their mammary organs are dispersed; their intestines are less voluminous than those of herbivorous animals, a provision, the object of which is to prevent the flesh which forms their food from putrefying, by remaining too long in the body.

*Order* 7.—*Cheiropterans (Bats; Vampyres;* and *Flying-cats).* The animals of this Order are distinguished by real organs for flight, formed of the skin extended between the legs, as described on a former occasion ;[1] their mammary organs, as in the Quadrumanes, are pectoral ; they are, in some points, connected with the flying opossum, flying squirrels, &c.

*Order* 8.—*Quadrumanes. (Monkeys; Apes; Baboons; Oran-outans.)* The great character that distinguishes this order is, a *moveable thumb* on their *lower* extremities *opposed* to the *fingers,* so that they can use the *carpus, metacarpus,* and *phalanges* of both extremities as *hands.* I have more than once had occasion to observe,[2] that certain tribes in the animal kingdom seem occasionally to form centres from which rays diverge towards different parts. The quadrumanes afford another example of this disposition in nature : the lory, for instance, looks towards the sloths ; the baboon, the *Cynocephalus* of the ancients, towards the dogs and bears ; the *aye aye,* amongst the Rodents, also might be taken

---

[1] See above, p. 156.

[2] See Vol. I. p. 275, and II. p. 20. 35.

for a quadrumane,[1] and several other instances occur.

*Sub-class* 1. *Order* 1.—The animals of this Order have puzzled Zoologists to ascertain their place and character. At first they were regarded as oviparous instead of mammiferous quadrupeds, and the *Ornythorhynchus* in particular, was thought to be something between bird and beast. The researches of Mr. Owen have almost proved that the animal just named does not leave the womb of its mother as an egg, requiring her incubation, to complete its birth; but in the form it is afterwards to maintain, in which case it must necessarily derive its support from her, by some lactescent organ, traces of which have been discovered. Its beak resembling that of a duck, and its webbed feet seem to connect it, in some degree, with the first Order of the *Birds;* but the entire scapular apparatus, the developement of the oviduct and uterus in both sides, the absence of the ligamentum teres, its four legs, and reptant motions, shew that it is most nearly connected with the *Reptiles.* The Echidna, by its extensile tongue, its food, and mode of taking it, approaches the anteaters: it also rolls itself up like an armadillo. The functions of the Order seem to be to keep in check the numbers of small animals; the

---

[1] See Vol. II. p. 210.

Echidna, the *ants;* and the Ornithorynchus,
which frequents the waters, some that are
*aquatic.* But we know very little of their habits
and history.

*Order* 2.—The animals of this Order are partly
herbivorous, and partly carnivorous. The wombat,[1] the koala,[2] the kanguroo,[3] and other New
Holland species, are herbivorous ; the phalangist[4]
of the Moluccas, lives upon the trees, and devours
insects as well as fruits. The New Holland
opossums[5] are very voracious, and devour carcasses as well as insects : they enter into the
houses, where their voracity is very troublesome.
That most common in America,[6] like the fox,
attacks poultry in the night, and sucks their
eggs. It is said to produce often sixteen young
ones in one litter, which, when first born, do not
weigh more than a *grain* each ! though blind
and almost shapeless, when placed in the pouch
they instinctively find the nipple, and adhere to
it till they attain the size of a mouse, which does
not take place till they are fifty days old, at which
period they begin to see ; after this they do not
wholly leave the pouch till they are as big as a
rat !! This statement is so extraordinary, that,
though apparently believed by Cuvier, on the
authority of Barton,[7] it seems almost incredible.

---

[1] *Phascolomys.*    [2] *Lipurus.*    [3] *Macropus.*
[4] *Phalangista orientalis.*    [5] *Dasyurus.*
[6] *Didelphis virginiana.*    [7] *Règn. An.* I. 176.

It is strange, as the animal seems common in America, that Say, or some other Zoologist of that country, has not turned his attention to it.

I have mentioned, on another occasion,[1] several particulars of the history of the kanguroo and koala, which I need not repeat here. Indeed our knowledge of the history and instincts of the Marsupian animals is very limited. Europe produces none. New Holland, some of the Asiatic islands, and North and South America, are their principal habitations. As the young of these animals leave the matrix of their mother at so early a period, and when, if they were exposed to the atmosphere, they must inevitably perish, it is evident that some such protection, as that with which Providence has furnished them, was necessary for the preservation of the race. Doubtless some wise and beneficial end is answered by the seeming premature nativity of these little creatures.

The opossums are peculiar to America, and are remarkable for having a greater number of teeth than any other animal, amounting in all to fifty; they approach the Quadrumanes, by having the thumb of their hind foot opposed to the fingers, whence they have been called Pedimanes, but it is not armed with a nail. They are usually stationed on the trees, where they pursue

---

[1] See above, p. 175, 211.

birds and insects, though, like the monkeys, they often eat fruit, and by this structure of the hind foot they can probably better support themselves on the branches. Many of the animals of this Order tend also to the *Rodents*, and others to the *Predaceans*.

*Sub-class* 2. *Order* 1.—At the foot of the present Class are found the most gigantic animals with which it has pleased God to people the globe that we inhabit.

The destruction, however, at least in the Arctic seas, of these animals, is so great, that it has been supposed, they are not suffered to live long enough to attain their full dimensions; but this has been doubted. Mr. Scoresby saw none in those seas that exceeded sixty-eight feet in length; but some are said to reach one hundred and twenty feet. I saw one, which was exhibited two years ago, in the King's Mews, the length of the skeleton of which was more than ninety feet. In the Antarctic seas, where the cupidity of mercantile enterprise does not occasion any great destruction of them, some are said even to reach the enormous length of one hundred and sixty feet. God has placed these Leviathans[1] where their enormous bulk can have full play, and their enormous appetite be fully satiated, in the vast and teeming depths of the

[1] See above, p. 432.

ocean, where, whether they move horizontally, or, by the aid of that powerful organ, their forked tail, seek the deep waters, there is space, and to spare, even for them.

The carnivorous, or predaceous Cetaceans may very conveniently be divided into sections by characters which distinguish their *maxillary* organs; the common whale,[1] and the fin-whale,[2] have their jaws armed with no real teeth, but only furnished with transverse plates, formed of what is called whalebone, consisting of a fibrous horny substance, sufficient for the mastication of their, for the most part, gelatinous food, which swarms in such infinite myriads in the Arctic and icy seas, that Scoresby calculates it would require eighty thousand persons, constantly employed from the Creation, to count the number of those existing simultaneously.

Animals of this section are further subdivided into those that have, and those that have not a dorsal fin. To the latter subdivision belongs the animal commonly distinguished as the *whale* by way of eminence,[3] and which is the principal object of the whale fishery. The senses of seeing and hearing in these animals, in the water, are extremely acute; and their eyes are so placed that they can see behind as well as before and above them, and for a great distance; but when

---

[1] *Balæna.*     [2] *Balænoptera.*     [3] *Balæna Mysitcetus.*

the head emerges from the water, this activity of sight and hearing ceases.

Their motions in the water are extremely rapid. They will sometimes assume a perpendicular position, with their head downwards, and rearing aloft their tremendous tail, lash the water with terrific violence, like the Indian god, churning the sea into foam, and filling the air with vapour. Sometimes by the motion of this organ, they produce a thundering noise. They will dive to the bottom of the ocean; and when confined in the shallows, these unwieldy monsters will sometimes leap out of the water. Their brain, compared with that of man, is very small. The weight of the brain of an adult man is often four pounds; that of a whale, nineteen feet long, only three pounds and a half; yet this is large compared with that of some other animals.

The *second* section of Cetaceans consists of those which have teeth only in their *upper* jaw. To this tribe belongs the *sea-unicorn*, or narwhal,[1] distinguished by its long tusk, or tusks, for there are sometimes two, extended in a horizontal direction.

To the *third* section belong those that have teeth only in their *lower* jaw : of this description are the spermaceti whales, or cachalots,[2] remark-

---

[1] *Monodon Monoceros.*    [2] *Physeter.*

able for their enormous head, sometimes occupying half the length of the body. Their teeth are long, and numerous, and all point outwards; opposite to them, in the upper jaw, is an equal number of cavities, in which the ends of the teeth are lodged, when the mouth is closed. These animals are said to grow sometimes to an enormous length; and to be very cruel and dangerous.

The *fourth* and last section of carnivorous Cetaceans consists of those that have teeth in both *upper* and *lower* jaws. To this the porpoise,[1] the grampus,[2] and the long celebrated dolphin[3] belong. These animals are more active than the preceding Cetaceans, and have a brain of greater volume. The common dolphin is gregarious, and remarkable for its frolicsome gambols, often foretelling a storm, during which they will leap entirely out of the water. They pursue and devour the gregarious migratory fishes, and will even eat offal and garbage. These animals, in their tooth-armed mouth, often opening wide, seem to exhibit some affinity to the aquatic Saurians, as has been remarked with regard to the Cetaceans in general.[4]

The end for which all these carnivorous Cetaceans were brought into existence by the Creator

---

[1] *Phocæna.*   [2] *Delphinus Orca.*
[3] *Delphinus Delphis.*   [4] See Vol. I. p. 30, 31.

of the universe, was evidently to keep within due limits, those animals, inhabitants of the northern and southern oceans, which were most given to increase, and which, were it not for some such check, might multiply to such a degree as would interfere with the general welfare.[1]

But the *vegetable* tenants of the ocean require to be kept within due limits, as well as the animal, amongst other creatures to whom this province is assigned, are some Cetaceans; thus preserving the general analogy observable in the animal Kingdom, which, in almost every Order, has its *cattle*, as well as its beasts of *prey*. Only three genera have been hitherto discovered to which this function is assigned, and all of them consisting of animals now in existence.

The Manatees,[2] belonging to this Sub-order, on account of their carrying their young with their flappers or fin-like legs, and their breasts, probably gave rise to the fable of the siren, or mermaid.

One of the most remarkable of the herbivorous Cetaceans, is the *Dugong*[3], which is the only animal yet known that grazes at the bottom of the sea usually in shallow inlets, which it is enabled to

---

[1] See Vol. I. p. 199—202.   [2] *Manatus Americanus.*
[3] *Halicore Dugong.*

accomplish by its power of suspending itself
steadily in the water, and by having its jaws
bent down at an angle, in such a manner as to
bring the mouth into nearly a vertical direction,
so that it can feed upon the sea-weeds much in
the same manner as a cow does upon the herbage.

Ruppel, a traveller in Africa, discovered a
second species of Dugong in the Red Sea; and
he is of opinion, that it was the skin of this
animal with which the Jews were commanded to
cover the tabernacle.[1]

*Order* 2. Whoever compares the genuine Pachyderms with the Cetaceans, will find many
points in which they resemble each other. As
the latter Order contains the largest *marine* animals, so does the former the giants that inhabit
the *earth*. With respect to their integument,
the skin of both is nearly naked, except in the
case of the swine, the daman,[2] the mammoth,
and some others; a very small eye characterizes
all, and a short tail; the blubber of the whale
seems to have its analogue in the fat that covers
the muscles of the swine. One of the most
remarkable animals of this Sub-order, is the
fossil one, which, on account of its enormous
tusks, is named *Deinotherium*.[3] It is found in

[1] *H. Tabernaculum.* See Exod. xxvi. 14. *Badgers'* skins in
our Translation.

[2] *Hyrax.*

[3] From the Gr. δεινος, terrible, and θηριον, wild beast.

the north of Europe, and specimens of its powerful jaws and tusks may be seen in the British Museum. From its lower jaw two powerful tusks rise as in the Hippopotamus, to which Mr. Owen regards it as approaching very near, and as forming the link that unites the Cetaceans to the Pachyderms. The herbivorous Cetaceans, in common with the generality of the Pachyderms, are likewise armed with tusks; so that the interval that separates the Hippopotamus and Deinotherium from the Dugong is not very wide.

The grand function of the, for the most part, mighty animals which constitute the tribe I am speaking of, seems to be that of inhabiting and finding their subsistence, in the tropical forests of the old world ; both Africa and Asia have each their own rhinoceros, and elephant, which, by their giant bulk, and irresistible strength, can make their way through the thickest forests or jungles. Even the swine, from the thickness of its skin, suffers nothing from pushing through bushes and underwood in search of acorns ; and most of these animals, by means of their tusks, muzzle, or horns, can dig up the roots that form their food. The hippopotamus seeks his provender in the African rivers, and by means of the tusks with which the under-jaw is armed,— in this differing from the dugong, in which the tusks are in the upper jaw,—is enabled to root up plants growing under the water. The tapir acts the same part nearly in the New World that the hippopotamus does in the old.

By the efforts of the Pachyderms, in general, in pursuit of their own means of subsistence, a way is often made for man more readily to traverse and turn to his purpose forests and woody districts, that would otherwise mock his efforts to penetrate into them. When we consider the vast bulk and armour of the rhinoceros, for instance, and the violence with which he endeavours to remove obstacles out of his path, we may in some degree calculate the momentum by which he is enabled to win his resistless way through the thickest and most entangled under-wood.

I need not enlarge on the *second* Sub-order of the Pachyderms, the Solipeds, the well-known equine and asinine tribes; every one must be struck by the contrast that their structure and characters exhibit to those of the *first* Sub-order, or typical ones. A fiery and intelligent eye; a *neck clothed with thunder*, to use the words of inspiration; a graceful form; speed that often outstrips the wind; are the distinctive characters which the highest tribe of them exhibits; while the other, though less beautiful, still has the organs of sight and hearing singularly conspicuous; a long tail; and its integument clothed with a shaggy coarse fur: besides these charac-

ters, the undivided hoof of both these tribes
forms also a most striking distinction. No animals, indeed, externally present characters more
diverse from each other than the soliped and
typical Pachyderms. God has given us these
animals, evidently, that we may employ them as
our *servants*, and their great function is, to carry
ourselves and our burthens; they also minister
in no small degree to our innocent pleasure and
amusements, as well as to our defence and
security.

*Order* 3.—Of all the different Orders of the
present Class, or indeed of all the Classes of animals, none are of so much importance to their
Lord as the *Ruminants*, which we are next to
consider; without them, hunger, cold, and
nakedness would beset him, or, at least, a large
portion of his comforts, with respect to articles
of food and clothing, must be cut off.

Cuvier divides this great Order into those that
have horns, and those that have none, and we
may here adopt his division, considering these
two sections as forming two Sub-orders. The
first of them, being the beasts of burden of more
than one nation, may be regarded as succeeding
the solipeds; these are the camels and drome-
daries, the lamas; and perhaps what is called
the musk-deer, also wanting horns, may be
placed amongst them. So that we have thus
before us animals that may be regarded as

looking towards the Solipeds, in the *camel* genus;
towards the sheep by its fleece, in the *lama;* and
towards the antelope tribes in the *musk.*

All the other Ruminants, the males at least,
are armed with two horns, either simple or
branching ; either hollow, or solid ; either persistent or deciduous. I feel disposed to consider
the giraffe, or camelopard, as an intermediate
form between the animals that are horned, and
those without horns, for its short, persistent,
solid horns, clothed with a velvet skin, seem
almost rudimentary. It may be regarded as
connecting, in some degree, the long necked
animals, the camel and lama, &c. with the deer
tribe.

These last, the most elegant and airy, both in
form and limb and motions, of the whole class,
placed in contrast with the clumsiness and bulk
of the Pachyderms, seem intended as one of
the principal ornaments of the globe we inhabit, and originally to be amongst the peculiar
favourites of its king and master man. Now,
instead of the innocuous animals, he takes into
his alliance, as his most intimate associates,
those that are best fitted to pursue and destroy,
as the dog, and the cheetah; and thus with the
help of the horse, he overtakes these beautiful
creatures, and, instead of caresses, they receive
death at his hands.

The head of these animals, in some, as the

rein-deer[1] in both sexes, but generally only in the males, is ornamented, as it were, with a branching forest,[2] formed by its antlers, or horns, which are solid, covered, as in the camelopard, with a velvet skin, but only during the period of growth, and annually deciduous; these are used by the males in their mutual combats. Amongst these light and airy animals, however, some of a larger and more robust stature are thus fitted for the use of man, as the rein-deer. The elk, or moose,[3] the wapiti[4] and red deers, emulate the horse in size, and are of great strength, though not yet employed by man.[5] Lastly, come the Ruminants, whose horns are hollow and naked, but persistent. To these belong the Antelopes, one species of which has four horns,[6] the goats, the sheep, and the bovine tribes. The species of the two last of these great families are particularly important to man, and are generally so well known as not to require to be treated of in detail. The bison,[7] with his shaggy mane, presents no slight analogy to the lion, the so called king of beasts; and the gnu, reckoned amongst the antelopes, seems to combine characters borrowed from the ox and the horse.

---

[1] *Cervus Tarandus.*   [2] French. *Bois.*   [3] *C. alces.*

[4] *C. strongyloceras.*   [5] See above, p. 184.

[6] *A. Chickara.*   [7] *Bos Urus.*

The function of this great Order of Ruminants, is not only to browse the herbage, and provide, by constantly trimming, and as it were mowing it, for its renewed verdure; many of them are employed also in pruning the trees, by feeding upon their branches; and there is not one that, in its place, does not contribute its part to the general welfare. The cattle *on a thousand hills* are distributed by their Great Creator according to certain laws, and by their actions in their several spheres, to promote certain ends, which neglected, or imperfectly provided for, would produce derangements that might affect a wide circumference.

*Order* 4.—Having, in a former part of the present volume, given an account of the principal tribes of this Order, I need not here do more than mention it, except by observing, that the members of it are principally inhabitants of the *new* world, the *Manis* and *Orycteropus*, being the only genera it contains that are found in the *old*.

*Order* 5.—The animals included in the Order of *Rodents*, or gnawers and nibblers, as I have before observed,[1] seem to occupy the same station amongst the Mammalians, that the *Hymenoptera* do amongst Insects, since they are the most

---

[1] See above, p. 209.

remarkable of any for the arts which Providence
has instructed them to exercise.   This, as well
as the preceding Order, seems very slightly
connected with the great tribe of Ruminants:
the Patagonian hare,[1] however, of the Pampas,
belonging to the Rodents, seems, in its light
and elegant form, to make the nearest approximation to that tribe.

Several of the animals of the Order before us
copy the members of the class of insects in one
of their most remarkable peculiarities; during
the cold or winter season, they become torpid.
This is the case with the dormouse,[2] the mar-
mots,[3] the prairie-dog,[4] and many other Rodents,
as well as with many predaceous Mammalians,
especially the insectivorous ones, as the hedge-
hogs.[5]   The mole, and the bats, and even some
of the largest animals, as the bear, are subject to
the same law.   When we consider the case of the
insectivorous animals of the present class, we see
at once the wisdom and goodness of the Lawgiver
in this enactment.   The reduction of the temperature, and other causes, have driven the insects
from the theatre they usually frequent, to remain
for a time without motion under the earth and

---

[1] *Cavia patagonica.*          [2] *Myoxus avellanarius.*
[3] *Arctomys.*
[4] *Spermophilus ludovicianus.    Faun. Boreal. Americ.* i. 156.
[5] *Erinaceus.*

other places of security, where they are safe from these their enemies; it was, therefore a kind and wise provision, that as their accustomed food was beyond their reach, they themselves should also be placed in a state not to require it. Many other animals amongst the Rodents, though they do not pass the winter in a state of absolute torpidity, retreat to what may be called their winter quarters, in which they have laid up a store of provisions against the evil days of winter. Of this description are many of the *murine* tribes, particularly the *hamster*,[1] which is furnished with a pouch on each side of its mouth, that it fills with grain to deposit in its burrow, for a winter store. Some will thus carry as much as three ounces at a time. The lemmings[2] also, whose destructive ravages I have before noticed,[3] especially that called the *economist*,[4] have similar habits, storing up roots instead of grain.

Generally speaking, it is the lowering of the temperature that induces Mammalians, as well as cold-blooded animals, to hybernate, and brings on a state of torpidity, or a cessation of the usual stimulus to locomotion and action: in which state, Mr. Owen remarks, *warm-blooded* animals become, as it were, *cold-blooded*. As a watch not wound up remains without motion,

[1] *Cricetus.*　　　[2] *Arvicola. Lemmus.*
[3] Vol. i. p. 91.　　[4] *L. œconomus.*

still retaining the power of resuming it, and when the mainspring recovers its elasticity is again enabled to act upon its wheels : so to animals *heat* is the key that winds up the wheels, and restores to the mainspring its powers of reaction. Hybernating animals have supernumerary cells, and generally become very fat in autumn, and it has been said that this fat supports them in their torpid state; it is found, however, that there has been but little of it consumed during the state of torpidity, but that it wastes very fast immediately after that state is ended. The Indians remark, with respect to the black bear, that it comes out in the spring with the same fat which it carries in in the autumn ; but after the exercise of only a few days it becomes lean.[1] A state of periodical rest may be necessary to the animals we are speaking of, not only as a means of protection from the effects of a low temperature, and on account of the impossibility of procuring their usual means of subsistence ; but since alternate rest and action are necessary to most animals, so a longer period of sleep may be required in some cases, by such cessation of action to keep the machine from wearing out too soon. Excess of heat we know produces the same effect as excess of cold, it disposes to sleep.[2] The tenrec,[3] a Madagascar

[1] *Fr. Boreal. Americ.* i. 20.
[2] *N. D. D'H. N.* xxxi. 387—390.      [3] *Seliger.*

animal, and the jerboa, fall into a kind of summer lethargy from that cause, which lasts some months.[1]

From the numerous instances of remarkable instincts exhibited by the animals of this Order, which might be selected, I must confine myself to one or two of the most singular. The hare is only noticed for its extreme timidity and watchfulness, and the rabbit for the burrows which it excavates for its own habitation, and as a nest for its young: but there is an animal related to them, the *rat-hare*,[2] which is gifted by by its Creator with a very singular instinct, on account of which it ought rather to be called the *hay-maker*, since man may or might have learned that part of the business of the agriculturist, which consists in providing a store of winter provender for his cattle, from this industrious animal. Professor Pallas was the first who described the quadruped exercising this remarkable function and gave an account of it. The Tungusians, who inhabit the country beyond the lake of Baikal, call it Pika, which has been adopted as its Trivial name.

These animals make their abode between the rocks, and during the summer employ themselves in making hay for a winter store. Inhabiting the most northern districts of the old world,

[1] *N. D. D'H. N.* xxxiii. 53.  [2] *Lagomys.*

the chain of Altaic Mountains, extending from Siberia to the confines of Asia and Kamtschatka,[1] they never appear in the plains, or in places exposed to observation ; but always select the rudest and most elevated spots, and often the centre of the most gloomy, and at the same time humid forests, where the herbage is fresh and abundant. They generally hollow out their burrows between the stones and in the clefts of the rocks, and sometimes in the holes of trees. Sometimes they live in solitude and sometimes in small societies, according to the nature of the mountains they inhabit.

About the middle of the month of August these little animals collect with admirable precaution their winter's provender, which is formed of select herbs, which they bring near their habitation and spread out to dry like hay. In September, they form heaps or stacks of the fodder they have collected under the rocks or in other places sheltered from the rain or snow. Where many of them have laboured together their stacks are sometimes as high as a man, and more than eight feet in diameter. A subterranean gallery leads from the burrow, below the mass of hay, so that neither frost nor snow can intercept their communication with it. Pal-

---

[1] Mr. Daines Barrington presented to the Royal Society an animal resembling the Pika found in Scotland, but probably a different species.

las had the patience to examine their provision
of hay piece by piece, and found it to consist
chiefly of the choicest grasses, and the sweetest
herbs, all cut when most vigorous, and dried so
slowly as to form a green and succulent fodder;
he found in it scarcely any ears, or blossoms,
or hard and woody stems, but some mixture of
bitter herbs, probably useful to render the rest
more wholesome.   These stacks of excellent
forage are sought out by the sable-hunters to
feed their harassed horses, and the (Jakutes)
natives of that part of Siberia, pilfer them, if I
may so call it, for the subsistence of their cattle.
Instead of imitating the foresight and industry
of the Pika, they rob it of its means of support,
and so devote the animals that set them so good
an example to famine and death.[1]   How much
better would it be if instead of robbing and
starving these interesting animals, they learned
from them to provide in the proper season a
supply of hay for the winter provender of their
horses.

But no animals in this, or indeed any other
Order of Mammalians, are so admirable for their
instincts and their results as the *beavers*.

I have more than once alluded to some
proceedings of these, seemingly, half-reasoning
animals, and shall now as briefly as possible

[1] *N. D. D'H. N.* xxvi. 407—410.

give some account of those fabricks in which
their wonderful instinct is principally manifested.
There are two writers who had great opportunities of gaining information concerning them;
Samuel Hearne, during his journey to the Northern Ocean, in the years 1769, 1770, 1771, and
1772; and Captain Cartwright, who resided
nearly sixteen years on the coast of Labrador.
To them I am principally indebted for the particulars of the history here given.

From the breaking up of the frost to the fall
of the leaf, the beavers desert their lodges, and
roam about unhoused, and unoccupied by their
usual labours, except that they have the foresight to begin felling their timber early in the
summer. They set about building some time in
the month of August. Those that erect their
habitations in small rivers or creeks, in which
the water is liable to be drained off, with wonderful sagacity provide against that evil by forming
a dike across the stream, almost straight where
the current is weak, but where it is more rapid,
curving more or less, with the convex side
opposed to the stream. They construct these
dikes or dams of the same materials as they do
their lodges, namely, of pieces of wood of any
kind, of stones, mud, and sand. These causeways oppose a sufficient barrier to the force, both
of water and ice; and as the willows, poplars, &c.
employed in constructing them, often strike root

in it, it becomes in time a green hedge, in which the birds build their nests. Cartwright says that he occasionally used them as bridges, but as they are level with the water, not without wetting his feet. By means of these erections the water is kept at a sufficient height, for it is absolutely necessary that there should be at least three feet of water above the extremity of the entry into their lodges, without which, in the hard frosts, it would be entirely closed. This entry is not on the land side, because such an opening might let in the wolverene, and other fierce beasts, but towards the water.

Cuvier, in his table above alluded to,[1] assigns only *four* pectoral teats to the female beaver; but Dr. Richardson states that she has *eight*, and the maximum of her young ones at eight or nine.[2] The number inhabiting one lodge seldom exceeds four old and six or eight young ones; the size of their houses, therefore, is regulated by the number of the family. Though built of the same materials, they are of much ruder structure than their causeways, and the only object of their erection appears to be a dry apartment to repose in, and where they can eat the food they occasionally get out of the water. It frequently happens, says Hearne, that some of the large houses have one or more partitions, but these are

---

[1] See above, p. 476.     [2] *Fn. Boreal. Amer.* i. p. 107.

merely part of the building left to support the
roof. He had seen one beaver lodge that had
nearly a dozen apartments under the same roof,
and, two or three excepted, none had any communication but by water. Cartwright says, that
when they build, their first step is to make choice
of a natural basin, of a certain depth, near the
bank where there is no rock; they then begin
to excavate under water, at the base of the bank,
which they enlarge upwards gradually, and so
as to form a declivity, till they reach the surface;
and of the earth which comes out of this cavity
they form a hillock, with which they mix small
pieces of wood, and even stones: they give this
hillock the form of a dome, from four to seven
feet high, from ten to twelve long, and from eight
to nine wide. As they proceed in heightening,
they hollow it out below, so as to form the lodge
which is to receive the family. At the anterior
part of this dwelling, they form a gentle declivity terminating at the water; so that they enter
and go out under water. The hunters name this
entrance the *angle*. The interior forms only a
single chamber resembling an oven. At a little
distance is the magazine for provisions. Here
they keep in store the roots of the yellow water-
lily, and the branches of the black spruce,[1] the
aspin,[2] and birch,[3] which they are careful to

---

[1] *Abies nigra.*     [2] *Populus tremula.*     [3] *Betula alba.*

plant in the mud. These form their subsistence. Their magazines sometimes contain a cart-load of these articles, and the beavers are so industrious, that they are always adding to their store.

There is a species of beaver found in the great rivers in Europe—the Danube, the Rhine, the Rhone, and the Weser, which has been regarded as synonymous with the beaver of Canada, but which, though it forms burrows or holes in the banks of those rivers which it frequents, does not, like them, erect any lodges, as above described. Does this instinct sleep in them, and require a certain degree of cold to awaken it, or are they a distinct species? Linné mentions one in Lapland, where the cold is sufficiently intense. Cuvier seems uncertain whether they ought to be considered as distinct. Beavers seem formerly to have existed in England; the town of Beverley (*Beaver-field*), in Yorkshire, seems to have taken its name from them, and its arms are three beavers.

Such are the principal operations that these wonderful animals, probably by the mixture of intellect with instinct, are instructed and adapted by their Creator to execute, that man, by studying them and their ways, may acknowledge the Power, Wisdom, and Goodness that formed and guides them.

The functions of the numerous tribes of this Order are various. The great majority may be said to be granivorous, or nucivorous, or even graminiverous; but many live upon dried vegetable substances, and wood. The *aye aye*, which approaches the Quadrumanes, appears to be insectivorous. Though many of them are great plagues to man, yet, by exciting his vigilance, they are useful to him, and they form the food of many of the lesser predaceous animals.

*Order* 6. The connection between the animals of which this Order consists, and the Rodents, seems not easily made out. The lowest tribe, the Amphibians, which Cuvier has placed immediately before the Marsupians, appears to have no connection with that Order, or any of the Rodents; and the morse, which forms his last genus of the tribe in question, appears evidently to look more towards the herbivorous cetaceans, the manatee,[1] &c. than to any other animals; the seals, indeed, may be regarded as tending towards the feline tribe. Amongst the other Predaceans, the hedgehog and tenrec present, I apprehend, something more than an analogy to the porcupines and some of the rats. The bear seems to look towards the sloth; and the feline race, in their whiskers and feet, look to the hares and rats.

The general functions of this Order are to check the tendency to increase not only in their

---

[1] *Cheiromys.*

own Class, the Mammalians, but in most of the other Classes of animals, more particularly those which man has taken into alliance with him, as cattle, and poultry, and game of every description. But where his action is greatest, theirs is usually least; and the most powerful devastators of the animal kingdom, the lions and the tigers, are found in the warmest climates, where nature is most prolific, and where man has not fully established his dominion, in the trackless and burning deserts of Libya, and in the impenetrable forests and jungles of India.

In more northern regions, the bears, the foxes, and other Mammalians, are employed in this department, though the former also eat roots and other vegetable substances,[1] and thus in the wild countries of the north supply the place of man, and keep the animal population under, and at a certain level, so that one may not encroach upon another. If the matter is closely investigated, we shall find that God has distributed and divided these predaceous animals to every country, in measure and momentum, as every one had need.

The necrophagous Mammalians[2] also, or those that devour dead carcases, such as the hyænas, dogs, and similar animals, are equally useful in removing infectious substances, which

[1] *Fn. Boreali Americ.* i. 15, 23, 28.    [2] *Carnivora.* Cuv.

in hot climates soon generate disease, and are always disgusting objects, and exercise a very important and beneficial function, devolved upon them by their Creator; for if all the animals exercising this function were removed from the earth, it would soon be depopulated, and a universal pestilence would destroy man, and all his subject animals.

*Order* 7. The animals of this Order, though evidently leading towards the Quadrumanes, seems less nearly connected with the insectivorous Predaceans of Cuvier, the hedgehog, mole, &c., and to approach nearer to some Marsupians, as the flying squirrel and the flying opossum. I therefore consider them as forming an Osculant Order, distinguished by their powers and organs of flight before sufficiently noticed.[1] They are nocturnal animals, and live entirely upon insects. In the winter, they become torpid, and suspend themselves by the claw of the thumb of the fore-foot, which is left free for this and other purposes.

*Order* 8. Linné evidently degraded *man* when he placed him in the same Order with the *monkey*, and even considered his genus *Homo* as consisting of two species, advancing the Ouran Outan[2] to the honour of being his congener, and a

---

[1] See above, p. 156.
[2] Written also *Ourang Outang*, and *Orang Otang*.

second species of man. Cuvier has, with great propriety, separated man, the heir of immortality, and *whose spirit goeth upward,* from the beast that perisheth, and *whose spirit goeth downward,*[1] and placed them in different Orders. Man has employed some animals in almost every Order, or taken them under his care; but there is only a single instance of a Quadrumane being so used. There is a kind of monkey,[2] a native of Madagascar, which, being of a gentle disposition, the natives of the southern part of that island take when they are young, and educate, as we do hounds, for the chase.[3]

The principal function of these animals is to live and move in the trees, amongst the branches in tropical countries, and they subsist upon fruits, roots, the eggs of birds, and insects. One object of their creation seems to be to hold the mirror to man, that he may see how ugly and disgusting an object he becomes when he gives himself up to vice and the slave of his passions. In fact, in every department of the animal kingdom, the moral instruction of his reasonable creature seems to have been one of the objects of Creative Wisdom: and the sloth and the glutton may be added to the mandril and baboon as equally calculated to cause him to view vice with

---

[1] Eccles. iii. 21.      [2] *Indris brevicaudatus.*
[3] *N. D. D'H. N.* xvi. 171.

disgust and abhorrence; as the bee, the ant, and the beaver, to excite him to industry, and prudence, and foresight; or the dove to peace and mutual love.

## CHAPTER XXV.

### *Functions and Instincts—Man.*

AFTER traversing the whole Animal Kingdom from its very lowest grades, and having arrived at Man, who confessedly stands at the head, and is the only *visible* king and lord of all the rest, it will be expected that I should devote a few pages to the world's master.

Baron Cuvier, with great propriety, places him by himself in a separate Order, distinguished from that which succeeds it, in his system, by the significant appellation of *Bimane*, indicating that his two hands are the instruments by which he subdues and governs the planet that he inhabits;[1] by which also he is enabled to embody his conceptions, and, as it were, to convert his thoughts into material subsistences.

I shall consider him both physically and metaphysically; physically, as to his actual *posi-*

---

[1] See above, p. 215.

*tion*, and as to his *action* upon his subjects and property, whether vegetable or animal; and metaphysically as to his connection with that world, to which his mind or spirit belongs. When I say that Man stands at the *head* of the creation, I do not mean to affirm that he combines in himself every physical attribute in perfection that is found in all the animals below him; for it is manifest to every one, that many of them far exceed him in the perfection of many of their organs, and in their qualities of various kinds. For *sight*, he cannot compete with the *eagle;* for *scent*, with the *hound*, or the *shark;* for *swiftness*, with the *roe-buck;* for strength and bulk, with the *elephant:* but it is in his *mind* that his superiority lies. There is in him a SPIRIT, an immaterial substance which constitutes him the sole representative here on earth, of the SPIRIT OF SPIRITS. He is the only member of the Animal Kingdom that partakes both of a heavenly and of an earthly nature,—that belongs both to a material and an immaterial world: and on this account it was that God, when he had created man, constituted him king over the whole sphere of animals with which he had peopled this globe that we inhabit. When his unhappy *fall* took place, the Divine Image was impaired, and consequently the dominion over those creatures, which formed a part of it, was proportionably weakened, and reduced to its present standard. But still, though weak-ened, it is not abrogated ; his subjects have not universally broken the yoke and burst the bonds of his dominion—a large portion of them still acknowledge him as their king and master ; and those that he has not subdued so as to make them do his bidding, still fear him and flee him : and even of these, there is none so fierce and intractable, that he has not found means to tame and subdue. And this is the position in which he now stands with respect to the animal king-dom ; he has that within him that enables him to master them, and apply such of them as are of a convertible nature, if I may so speak, to work his will and answer his purpose.

The functions of man, with regard to the world in which he is now placed, are all in-cluded in his *action* upon the sphere of animals and vegetables, and in their *re-action* upon him. If we survey all nature, wherever we turn our eyes, or wherever we direct our thoughts, we see the action of antagonist powers, a flux and reflux, by which the Great Builder of the universe supports the vast machine, and main-tains all the motions that he has generated in it. The same principle is at work in every descrip-tion of beings in our own planet ; every action of man upon any object of the world, without him, produces a reaction from that object, at-tended often by important results.

The action of man upon the world without him, is *threefold*. His *first* action upon them is, that of the mind to contemplate them, so as to gain a knowledge of their forms and structure—of their habits and instincts—of their meaning and uses. His *second* action upon them, having studied their natures, and discovered how they may be made profitable to him, is to collect and multiply such species as he finds will, in any way, answer his purpose. His *third* action upon them is to diminish and keep within due limits those species that experience teaches him are noxious and prejudicial either to himself, or those animals that he has taken into alliance with him, which are principal sources of wealth to him, and minister to his daily use, comfort, and enjoyment.

If we consider the predaceous animals, we shall find in them a greater tendency to multiply than in those that content themselves with grazing the herbage ; they generally produce more young at a birth ; and their period of gestation is often shorter, so as to admit of more than one litter in the year ; so that, unless some means were used to reduce their numbers within a certain limit, the whole race of herbivorous animals must perish. Hence arose the first kind of *war*. Man armed himself to destroy such of his subjects as had rejected his dominion, and even contended with him for the possession of the earth, and to have license to devour at will its more peaceful inhabitants. A similar cause generated the other and more fearful kind of war, of man with man. *Whence come wars and fightings amongst you*, saith the Apostle ;[1] *come they not hence, even of your lusts that war in your members?*

The highest view that we can take of man is that which looks upon him as belonging to a spiritual as well as a material world. The end of the creation of the earth, says the father and founder of Natural History, is the glory of God, from the works of nature, by man only.[2] And, as the same pious author observes, " How contemptible is man," if he does not aim at this end of his creation, if he does not strive to raise himself above the low pursuits that usually occupy his mind ![3] The heavens indeed declare the glory of God, and the firmament sheweth the work of his hands. Day unto day uttereth speech, and night unto night sheweth knowledge.[4] The beasts of the field honour him, and all creatures that he hath made glorify him. But man must study the book open before him ; and the more he studies it, the more audible to him will be the general voice to his spiritual ear,

[1] James, iv. 1.

[2] *Finis creationis telluris est gloria Dei exopere naturæ per hominem solum.* Linn. *Syst. Nat.* i. *Introit.* i.

[3] *O quam contempta res est homæ nisi supra humana se erexerit.* Ibid.

[4] Ps. xix. 1, 2.

and he will clearly perceive that every created thing glorifies God in its place, by fulfilling his will, and the great purpose of his providence ; but that he himself alone can give a tongue to every creature, and pronounce for all a general doxology.

But further, in contemplating them, he will not only behold the glory of the Godhead reflected, but, from their several instincts and characters, he may derive much spiritual instruction. Whoever surveys the three kingdoms of nature with any attention, will discover in every department objects that, without any affinity, appear to represent each other. Thus we have minerals that, under certain circumstances, as it were, vegetate, and shoot into various forms, representing trees and plants : there are plants that represent insects, and, vice versa, insects that simulate plants ; and the Zoophytes have received their name from this resemblance.[1] And as we ascend the scale, every where a series of references of one thing to another may be traced, so as to render it very probable that every created thing has its representative somewhere in nature. Nor is this resemblance confined to *forms ;* it extends also to *character.* If we begin at the bottom of the scale, and ascend up to man, we shall find *two* descriptions in almost every class, and even tribe of animals : one,

[1] Vol. I. p. 149, 156, 169.

ferocious in their aspect, often rapid in their motions, predaceous in their habits, preying upon their fellows, and living by rapine and bloodshed; while the other is quiet and harmless, making no attacks, shedding no blood, and subsisting mostly on a vegetable diet.

Since God created nothing in vain, we may rest assured that this system of *representation* was established with a particular view. The most common mode of instruction is placing certain signs or symbols before the eye of the learner, which represent sounds or ideas; and so the great Instructor of man placed this world before him as an open though mystical book, in which the different objects were the letters and words of a language, from the study of which he might gain wisdom of various kinds, and be instructed in such truths relating to that spiritual world, to which his soul belonged, as God saw fit thus to reveal to him. In the first place, by observing that one object in nature represented another, he would be taught that all things are significant, as well as intended to act a certain part in the general drama; and further, as he proceeded to trace the analogies of character, in its two great branches just alluded to upwards, he would be led to the knowledge of the doctrine thus symbolically revealed—that in the invisible world there are two classes of spirits—one benevolent and beneficent, and the other malevolent and mischievous; characters which, after his fall, he would find even exemplified in individuals of his own species.

But after the unhappy fall of man, this mode of instruction by natural and other objects used symbolically, though it pervades the whole law of Moses, and the writings of the prophets, as well as several parts of the New Testament, gradually gave place to the clearer light of a Revelation, not by symbols, but by the words and language of man, which *he that runs may* often *read;* yet still it is a very useful and interesting study, and belongs to man as the principal inhabitant of a world stored with symbols, to ascertain what God intended to signify by the objects that he has created and placed before him, as well as to know their natures and uses. When we recollect what the Apostle tells us, that *the invisible things* of God *from the creation of the world are clearly seen, being understood by the things that are made,*[1] and that spiritual truths are reflected as by a mirror, and shewn, as it were, enigmatically,[2] we shall be convinced that, in this view, the study of nature, if properly conducted, may be made of the first importance.

In this enumeration and history of the prin-

---

[1] *Rom.* i. 20.      [2] 1 *Cor.* xiii. 12.

cipal tribes of the Animal Kingdom, we have traced in every page the footsteps of infinite Wisdom, Power, and Goodness. In our ascent from the most minute and least animated parts of that Kingdom to man himself, we have seen in every department that nothing was left to chance, or the rule of circumstances, but every thing was adapted by its structure and organization for the situation in which it was to be placed, and the functions it was to discharge; that though every being, or group of beings, had separate interests, and wants, all were made to subserve to a common purpose, and to promote a common object; and that though there was a general and unceasing conflict between the members of this sphere of beings, introducing apparently death and destruction into every part of it, yet that by this great mass of seeming evil pervading the whole circuit of the animal creation, the renewed health and vigour of the entire system was maintained. A part suffers for the benefit and salvation of the whole; so that the doctrine of the sufferings of one creature, by the will of God, being necessary to promote the welfare of another, is irrefragably established by every thing we see in nature; and further, that there is an unseen hand directing all to accomplish this great object, and taking care that the destruction shall in no case exceed the necessity.

Well, then, may all finally exclaim, in the words of the Divine Psalmist :—

*O Lord, how manifold are thy works, in WISDOM hast thou made them all; the earth is full of thy riches;*

*So is the great and wide sea also, wherein are things creeping innumerable both small and great beasts.*

*These wait all upon thee: that thou mayest give them meat in due season.*

*When thou givest them they gather it: and when thou openest thy hand they are filled with good.*

*When thou hidest thy face they are troubled: when thou takest away their breath they die, and are turned again to their dust.*

*When thou lettest thy breath go forth they shall be made: and thou shalt renew the face of the earth.*

# INDEX.

FINIS.

C. WHITTINGHAM, TOOKS COURT, CHANCERY LANE.

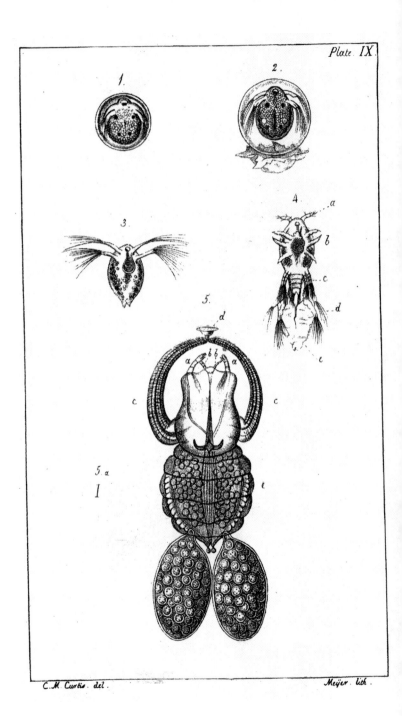

Plate IX

1.

2.

3.

4.

5.

5. a.

I

C.M. Curtis. del.

Meijer. lith.

Plate X.

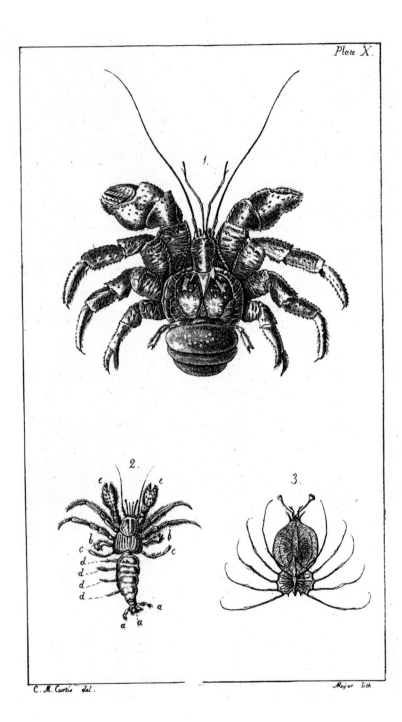

C. M. Curtis del.

Meijer lith.

Plate XI.

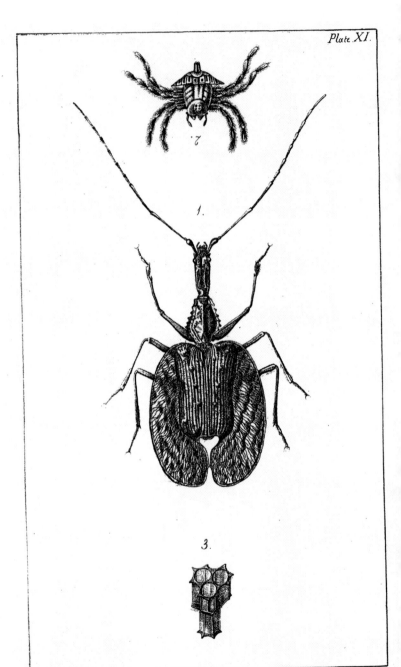

C. M. Curtis del.

Mayer lith.

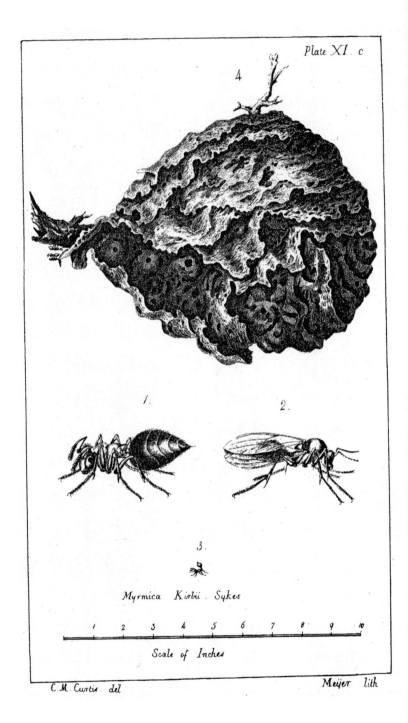

Plate XI. c

4

1.

2.

3.

Myrmica Kirbii Sykes

Scale of Inches

C.M. Curtis del

Meijer lith

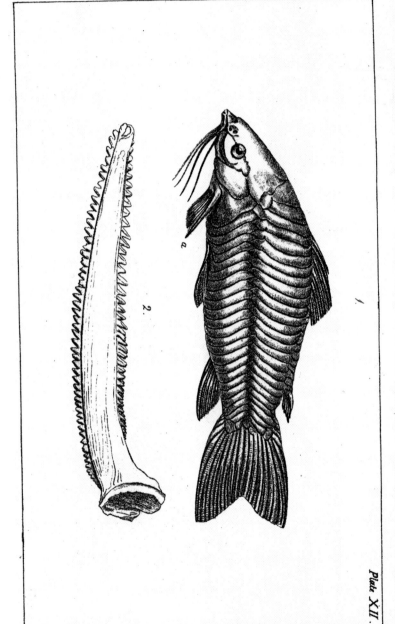

Plate XII.

C. M. Curtis. del.

Meijer. lith.

Plate XIII.

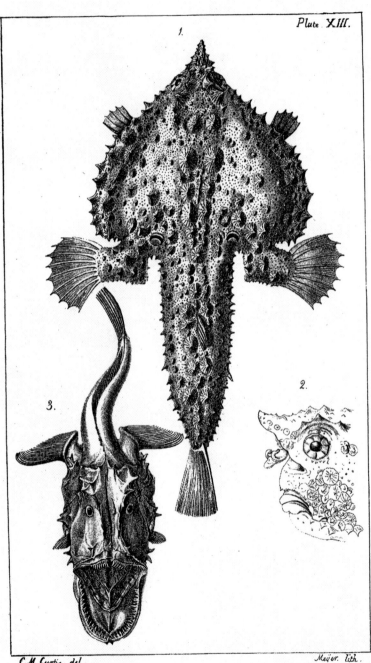

C.M. Curtis. del.

Meijer. lith.

Plate XIV.

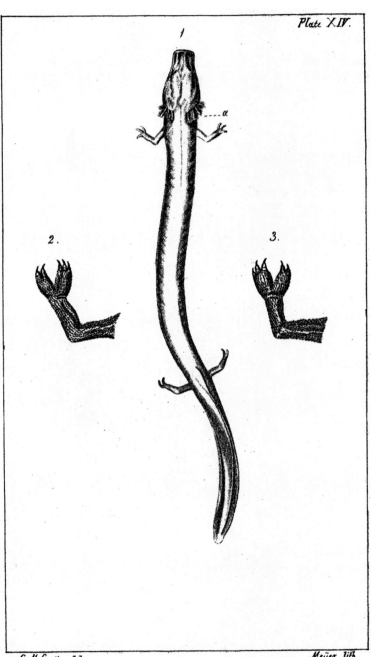

C M Curtis. del.

Meijer. lith.

Plate XVII

107940